James Mudge

Honey from many Hives

James Mudge

Honey from many Hives

ISBN/EAN: 9783743324534

Manufactured in Europe, USA, Canada, Australia, Japa

Cover: Foto ©berggeist007 / pixelio.de

Manufactured and distributed by brebook publishing software
(www.brebook.com)

James Mudge

Honey from many Hives

HONEY

FROM MANY HIVES

GATHERED BY

REV. JAMES MUDGE, D.D.

NEW YORK: EATON & MAINS

CINCINNATI: CURTS & JENNINGS

Eaton & Mains Press,
150 Fifth Avenue, New York.

TO THE

MEMORY OF MY MOTHER,

AND THE MANY OTHER GOOD WOMEN AND MEN

WHO HAVE HELPED ME IN

HOLY LIVING.

SALUTATORY.

THIS book, O reader, is for your closet, your secret place of private prayer and meditation. Such a place I trust you have. Busy times, to be sure, are these, and much seems to press upon us for doing; but what shall it profit if we gain all knowledge and all riches, and even cast out many of the devils that torment the age, while we do not properly know our own souls or make any real acquaintance with God? Take things a little more calmly. It needs time to be holy. Give ten minutes a day to quietly pondering some of the paragraphs which you will find in these carefully culled pages. Such a practice will work the most beneficent of revolutions in your life. For before you have penetrated far into this volume you will discover that it is not like other books. It contains the cream of many centuries, and could in no way have been produced by any one man, however wise or saintly. If you shall learn to love it and prize it at its true worth, you will make it your close and constant companion, nor will you consent to part with it for many times its price. Take it, then, not only into your closet, but into your mind and heart, and become by means of it a full sharer in the sacred joys of those who walk with God.

Natick, Mass. J. M.

CONTENTS.

HONEY FROM MANY HIVES.

DEVOTIONAL READING.

Of reading in general it may justly be said that he who has a taste for it has greater riches than the treasures of India. Truly happy is the man who has thoroughly learned how to eat paper and drink ink; that is, how to turn to best account the stores of learning that are wrapped up in printed volumes. The lover of books has an unfailing resource. Rainy days do not damp his enjoyment. Neither the heat of summer nor the cold of winter materially interferes with his delightful occupation. The loss of friends does not leave him friendless. He can make new acquaintances even in old age, and he can at any time renew his intercourse with those that were dear to him long ago. God be thanked for books—purveyors of information, stimulators of thought, unfailing entertainers, the tools of those who work in the realm of mind, the true levelers, giving easy access to the most select society. They are indeed "lighthouses erected in the great sea of time," throwing their effulgence over coasts and waves that without them would be full of danger to the mariner. They are comfortable inns established along the thoroughfares of life for the refreshment and solace of the weary traveler.

By devotional reading we mean the perusal of such books as are adapted to aid the spiritual life. Its value may be shown in various ways. For it has close connection with almost all the means of grace. Nearly all the processes of Christian growth are more or less vitally allied to it. Take, for example, prayer. Prayer is the grand difficulty with most souls, though they do not generally know it. The reason why they do not go forward is because they do not really and effectively pray. They know not what to ask for, both their own needs and the divine provisions being very largely hidden from them, and they do very little genuine asking. They keep up the practice to satisfy their conscience, but their petitions are formal, routine, unhelpful affairs which do not bring them into inspiring communion with God. The lamp of prayer, one may say, burns dim, and is often almost at the point of extinction. And where such is the case it is no wonder that the religious life is feeble. The remedy is to pour in oil. Very frequently do the old writers use this figure of speech to indicate the relation between devotional reading and fruitful supplication. The pertinency of it is evident. As the literal flame expires without food, so will the spiritual. When the mind has been sucked dry of uplifting thoughts by the multiplicity of distracting temporal interests that continually prey upon it, a fresh supply must be provided. When the attention has been long engrossed by earthly objects that thrust themselves persistently, and perhaps legitimately, into the mind, taking for a season full possession of the current of reflection, outside aid is required to turn that current successfully into another channel. Spiritual reading is just the thing. It invigorates the intellect, refreshes the emotions, and through them reaches the will. It has an invaluable power of suggestiveness. The affections are stirred. The cold heart is warmed. The laggard purpose is quickened. There is a general arousement of the whole soul. Now one can pray. He feels ashamed that he has fallen so far behind the examples of which he reads. He learns what his real needs are, and how best to meet them. Divine impulses leap into his heart from off the printed page. God speaks to him through the pens of his choicest children. Acts of faith, hope, love, and desire become easy. He takes a new start. His whole life becomes pitched on a higher key, and the process of celestial transformation is greatly accelerated.

This kind of reading is not only oil for the lamp of prayer, but bread and meat which may be turned into strength for Christian activity. That mind which is largely ignorant of the devices of Satan is not properly fortified against temptation so as to readily repel it. And these devices are so multifarious that something more than personal experience is necessary to make one fully acquainted with them. To wait for such experience would mean a sad loss of time and waste of opportunity. One might as well insist on learning the art of war solely by one's own battles. The wiser way is to draw on the stores of the past, utilizing the experience and observation of others, and thus avoid repeating their mistakes. He is best qualified to pull down the strongholds of the enemy, and to rout his forces in the campaign, or capture his country, who has been a diligent student of all other campaigns. This, at least, greatly helps. And idleness or weakness on the part of the Christian warrior will be far less common when he has become thoroughly familiar with the successes of his comrades. He will be stimulated as well as instructed by what they have done; less likely to yield to indolence, better qualified to win victory.

Very few realize how important for the proper advancement of spirituality is the cultivation of a taste for reading. A master in these things has put on record his estimate that he who begins a devout life without such a taste may consider the ordinary difficulties multiplied in his case by at least ten. However accurate this may be, it is clear that such a person is at a very great disadvantage. All the best writers are agreed on this point. He is not likely to be very thoughtful. He will fall into many errors which otherwise might have been easily avoided. He will be ignorant of those best methods which the wisdom of the ages has brought out. That which holy and learned men have by long contemplations received from God, and which he might with very little labor make his own, he will not know, and the lack of that knowledge will plunge him into many difficulties. He will blunder and stumble along where he might have run or soared.

To be sure, one may learn much by word of mouth. The pulpit is appointed in part for this very thing, that the man of God, or the man of godly tendencies, may be thoroughly furnished unto good works. But good teachers are rare. And though preaching of some sort or other is nearly always accessible, it is by no means always of the sort most suitable to promote sound and rapid growth in grace. But in the right kind of a book, procurable now for a very small sum, one has a preacher continually at hand. He is not confined to a special day or place. He may be returned to again and again, may be heard and reheard when one is most at leisure or most in need. Moreover, he speaks boldly, and with no danger of personal offense, what no individual would dare tell us to our face. He pricks us in our tenderest points, and lays bare the hideousness of our darling sins. This is a great advantage. It is a benefit, also, that we can take a little at a time, as we are able to bear it.

Spiritual reading, then, it is scarcely too much to

say, has in these modern times, and especially in so enlightened a land as America, reached a dignity and a consequence that puts it nearly on a level, for Christian people, with the listening to sermons. May it not be properly affirmed that the reading of religious books should now be regarded in the light of a truly divine ordinance? Has not literature come to be one of the most effective forms of preaching? Surely preaching, which all recognize as ordained of God, should not be restricted in its meaning to the delivering of a set discourse in a house of worship. If it be taken in the somewhat broader sense of the communication of divine truth to men through human instrumentality, then it will certainly include the use of the pen as well as the lips, and reading will be as much a duty as hearing. One may hear in the closet with the inward ear as well as in the church with the outward ear. Ought there not to be the same solemnity and sense of obligation in the one case as in the other?

Who will deny that if bad books have a mighty influence for evil, as we continually note with loud lamentation, good ones may and must be laid hold of for blessings? If "a companion of fools shall be destroyed," "he that walketh with wise men shall be wise;" and such walking is nowhere easier than in a little corner with a little book. Most certainly we need all the help we can get for making headway against the demoralizing tendencies of the day. To neglect the aid offered by some inspiring manual of devotion in the shape of a well-written biography, a series of confidential letters, a collection of hymns, or a treatise on the highest possibilities of grace, to the value of which aid such multitudes all down the ages bear ready testimony, is to falsify our profession of strong desire for the fullness of God; is to expect the end without the use of the means, and to prepare for ourselves disappointment and at least comparative failure.

But to be greatly impressed with the importance of good books is one thing; to know how to use them is quite another. A few counsels may be in place.

It is not best to fly too fast from flower to flower. A leisurely process is most beneficial. There must be time to ruminate and digest. The gentle showers are the ones that soak into the earth and fructify the vegetation. So one must bend over a good book with calm attention, quiet appreciation, and much meditation. As the birds stop when they drink a little and lift their eyes to heaven, one may read a few sentences and then turn them into prayer, looking up for help to comprehend and practice. A single sentence taken into the mind and thoroughly turned over there, till its whole bearing and application to daily life is clearly seen, is worth more than whole pages cursorily perused. Not many have sufficient wisdom to see that to go slow is often the quickest means of reaching the desired end. A solid truth really made one's own is a permanent acquisition. And when in course of time many such gains are securely harvested the character is wonderfully enriched. To read merely from curiosity, or for purposes of controversy, or because one feels that it is the proper thing to do, is a very different thing from reading with a single eye to personal improvement and an eager desire for advancement in goodness. He who pursues the latter course, reading for himself rather than for others, is the one who will make most progress.

It is an admirable plan to read with pen or pencil in hand. If the book is one's own, its margin may well be filled with neatly written comments and reflections. If it has been borrowed, then there should be transcription of its best expressions. Indeed, it is very desirable that each should construct a manuscript volume for himself. It need not be large, but it will surely become very precious. Into this volume should go certain passages of Scripture that have been proved and tried; texts that throb with life and flame with light; stanzas of hymns and parts of religious poems that have in them a mighty power of inspiration; precepts and proverbs and mottoes and maxims that seem to condense the wisdom of many centuries and yet have personal relation to one's own position; morning meditations, birthday resolutions, Sunday reflections, and, in short, the choice result of one's best moments. Such a volume will be the history of one's inner life. It will hold a record of hilltop experiences, where from special mounts of vision God showed one the wonders of the Canaan land or revealed how the temple of character should be built. It is no small recommendation of this practice that it was followed by seraphic John Fletcher, one of the most holy men that has ever blessed the earth, towering high above the generality of Christians, and enjoying closest fellowship with God. There is still in existence (held in safe and reverent keeping for more than a century past) a small, square book, strongly bound in leather, and containing about two hundred closely written pages, which was his closet companion. With its thoughts and rules he nourished his soul in private. With its spiritual exercises and disciplinary regulations, its tests and standards of self-examination, he sought to perfect himself in the love of God and in the minutest details of conduct. One feels, as he looks into this little manual of devotion which was so dear to the saint, that he is almost watching the way in which that saintliness was evolved. The lovely growth of goodness had at its root the patient discipline here outlined and portrayed. Here is the workshop from which the finished product was at last brought forth. It was mainly prepared when he was about twenty-seven years of age, although no doubt it grew considerably in the days subsequent to that period. We see no reason why a book like this should not be constructed by everyone who is in dead earnest to be all the Lord's, and so is ready to lay hold of every available means that gives promise of assistance in the mighty undertaking. Attention is thus concentrated, thought is clarified, mind and heart are kept on the alert, and much of permanent value is preserved which would otherwise vanish with the hurrying years. No one who has not tried it can fully realize what an aid pen and paper may become in furthering religious advancement.

It is well to have a fixed time in the day for devotional reading. Some have formed the habit of reading a little in connection with all their closet seasons. And these seasons have been observed, when nothing unusual occurred to disturb the routine, morning, noon, and night. In addition to prayer, Scripture, and perhaps a hymn, they have prized a paragraph from some good book, keeping a number on hand; varying the selection to suit the time of day—part of a sermon, perhaps, in the morning, a Bible comment at midday, a biography in the evening. Those who have some leisure and, what is still more essential, great zeal, can readily accomplish this. Others will find, probably, a single period for this kind of reading all that they can with regularity compass. Let the period be selected when there will be least interruption. With some it will be directly after dinner, with others immediately before retiring, while still others find that by rising at an earlier hour than would otherwise be necessary they can give their freshest powers to God and make the very best possible preparation for the work before them. But, whatever the time chosen, regularity is essential to the largest results. If only a few moments can be secured, so that not more than a page a day can be read, even that, with the extra opportunity of the Sundays, will mean, if it be continuously kept up, several volumes a year.

One should become acquainted with the standard works. Undoubtedly there are excellent books dropping from the press year by year, and among them we may sometimes discover that particular production which has a special message for us, wonderfully adapted to our peculiar need, written, as it were, for our eye and heart. We should be on the lookout for such a prize and purchase it promptly. Nevertheless, there are certain volumes which have been so long fed upon by the Church, which have survived so many vicissitudes of time, as to create a strong presumption in their favor. We may naturally expect that what many generations of Christ's children have drawn profit from will prove also profitable to us, and hence we approach such writings with large expectations that are not often disappointed. We shall not, of course, find them all equally suited to our own times or our own individual tastes. But it will be very strange if some of them do not become exceedingly dear tó us, veritable wells of salvation out of which we shall draw water with great delight. Opinions, no doubt, will differ considerably as to just what these standard works may be. No two persons would make precisely the same list, if it were of any length. A few, however, are in all men's mouths, and quite a number of others would obtain the general suffrage as standing in the front rank. But it is very noticeable, and indeed inevitable, that nearly all these ancient books, being written for a time so far separated from our own, in a foreign country most probably, and perhaps by an author belonging to a different branch of the Christian Church, contain very much that is not adapted to these days or our circumstances. To much of it, very likely, exception must be taken. Only a small part will be suitable for prolonged meditation, and fit to be implicitly followed. But that small part will be of priceless worth.

It was a consideration of this fact which led to the preparation of the book which the reader now holds in his hand. Some forty volumes, most of them such as would be accounted among the classics in this department, have been carefully searched and made to yield their choicest, most distinctive thoughts on the great fundamental themes most closely associated with devout living. Thus, within a small compass and at a moderate price, the cream of these twoscore books, some of them rare, is brought within the reach of everyone whose tastes run in this direction. It is not to be supposed that all would make exactly the same selections as those here presented, and much has had to be left out which would be of profit; but it is believed that every paragraph of this book will richly repay repeated perusal, and that in every case there is given a fair sample of the original author's best contribution to the thought of the world. There can be no question that he who properly masters these thoughts and follows these precepts will achieve a splendid character and possess a happiness such as few of earth's millions can at all conceive. It is hoped that this little book will be found worthy to lie on many a table beside the Bible and the Hymnal, as in full harmony with their teachings and as containing the best available collection of uninspired prose homilies on holiness. If it shall be made as great a means of grace to those who read it as its preparation, extending over many years, has been made to him who has compiled it, much good will certainly be done and many hearts give fresh glory to God through all eternity. May he grant it, for his name's sake! Amen.

"THE IMITATION OF CHRIST."

THE *De Imitatione Christi*—for such is its title in the original Latin—is so well known to all readers of good books that it hardly needs much introduction. It easily stands at the head of its class. Among uninspired volumes it ranks first for diffusion and popularity. Its editions in various lands, languages, and ages are to be counted by the thousand. Many years ago no less than sixty translations were known to have been made from it into modern tongues, and the number must be now much increased.

Its reputed writer is Thomas Hamerken, commonly called à Kempis from a little town near Cologne, where he was born in 1380. It is somewhat doubtful whether he wrote it; the dispute about the matter has filled a hundred volumes, and many are inclined to ascribe it to John Gerson, Chancellor of the University of Paris, who lived from 1363 to 1424. But it is probable that Thomas à Kempis will always retain the credit of the authorship. Not much is known about his life save that he spent seventy-one of the ninety-one years to which it was extended in the monastery of St. Agnes in the diocese of Cologne, of which he rose to be subprior, or, as some say, superior. Quiet industry in book copying, preaching, composing treatises, and other such exercises, together with lonely contemplation and secret prayer, filled up the gently gliding days; and a volume was produced (among others of inferior merit) which all the devout have agreed to put in the first place among religious manuals.

Dean Farrar has well said: "It is the legacy of the ages, it is the gospel of monasticism, it is the psalter of the solitary, it is the cyclic utterance of the mystic, it is the epic poem of the inner life. Whoever was the composer of the book did but gather into one rich casket the religious workings and interior consolations, the wisdom of the solitary experiences which had been wrung from many ages of Christian life." It was one of the important elements in the making of John Wesley, as in the case of multitudes more. During that formative period at Oxford, when he was laying out the lines on which his life was to be guided, Dr. Abel Stevens says, "he pored over the pages of that marvelous book, *De Imitatione Christi*, which has lent the fragrance of its sanctity to every language of the civilized world, and which by its peculiar appositeness to almost every aspiration, misgiving, or consolation of devout minds, has seemed more a production of divine inspiration than any other work in Christian literature except the Scriptures. It had been a favorite with his father, his 'great and old companion.'" After reading à Kempis Wesley says:

"I saw that simplicity of intention and purity of affection, one design in all we speak or do, one desire ruling all our tempers, are indeed the wings of the soul, without which she can never ascend to God. I sought after this from that hour." So grateful was he for the help afforded him by the book, and so highly did he prize it, that just as soon as he was in a position to use the printing press he translated it for his people and published it in an abridged form, calling it *The Christian's Pattern,* under which title the Methodist Book Concern still issues it.

Being written nearly five hundred years ago, and in a monastery, the book is, of course, not altogether adapted in every section to our greatly different modern life. But its main principles are perennial, and some of its sentences are very searching.

The quotations which we append contain, we believe, the very best portions; but there is the less need that we make extensive selection since the whole book is easily accessible in cheap and convenient forms. We advise the reader to procure a copy, and choose for repeated perusal those particular parts best adapted to his individual wants.

ZEAL FOR IMPROVEMENT.

The principal obstacle to the reformation and improvement of life is dread of the difficulty and labor of the contest. Only they make eminent advances in holiness who resolutely endeavor to conquer in those things that are most disagreeable and most opposite to their appetites and desires; and then chiefly does a man most advance to higher degrees of the grace of God, when he most overcomes himself, and most mortifies his own spirit.

But though all men have not the same degree of evil to overcome, yet a diligent Christian, zealous of good works, who has more and stronger passions to subdue, will be able to make a greater progress than he that is inwardly calm and outwardly regular, but less fervent in the pursuit of holiness.

Two things are highly useful to perfect amendment: to withdraw from those sinful gratifications to which nature is most inclined, and to labor after that virtue in which we are most deficient. Be particularly careful, also, to avoid those tempers and actions that chiefly and most frequently displease thee in others. Wherever thou art, turn everything to an occasion of improvement: if thou behold or hear of good examples, let them kindle in thee an ardent desire of imitation; if thou seest anything blamable, beware of doing it thyself; or if thou hast done it, endeavor to amend it the sooner. The zealous and watchful Christian bears patiently and performs cheerfully whatever is commanded; but he that is cold and negligent suffers tribulation upon tribulation, and of all men is most miserable; for he is destitute of inward and spiritual comfort, and to that which is outward and carnal he is forbidden to have recourse.

When a man is so far advanced in the Christian life as not to seek consolation from any created thing, then does he first begin perfectly to enjoy God; then "in whatever state he is, he will therewith be content;" then neither can prosperity exalt nor adversity depress him; but his heart is wholly fixed and established in God, who is his All in All. Reflect that it is only the fervent and diligent soul that is prepared for all duty and all events; that it is greater toil to resist evil habits and violent passions than to sweat at the hardest labor; that he who is not careful to resist and subdue small sins will insensibly fall into greater, and that thou shalt always have joy in the evening if thou hast spent the day well. Watch over thyself, therefore; excite and admonish thyself, and, whatever is done by others, do not neglect thyself. Thou wilt make advances in imitating the life of Christ in proportion to the violence with which thou deniest thyself.

TRUE LEARNING.

He is truly good who hath great charity; he is truly great who is little in his own estimation and rates at nothing the summit of worldly honor; he is truly wise who "counts all earthly things as dross that he may win Christ;" and he is truly learned who hath learned to do the will of God.

There is no other cause of perplexity and disquiet but an unsubdued will and unmortified affections. A holy and spiritual mind becomes the master of all his outward acts; he does not suffer himself to be led by them to the indulgence of inordinate affections that terminate in self, but subjects them to the unalterable judgment of an illuminated and sanctified spirit.

No conflict is so severe as his who labors to subdue himself; but in this we must be continually engaged if we would be strengthened in the inner man and make real progress toward perfection. Indeed, the highest perfection we can attain to in the present state is alloyed with much imperfection, and our best knowledge is obscured by the shades of ignorance. Because men are more solicitous to learn much than to live well they fall into error, and receive little or no benefit from their studies. Assuredly in the approaching day of universal judgment it will not be inquired what we have read, but what we have done; not how eloquently we have spoken, but how holily we have lived.

RIGHT DESIRES.

Let this be the language of all thy requests: Lord, if it be pleasing to thee, may this be granted or withheld; Lord, if this tend to thy honor, let it be done in thy name. If thou seest that this is expedient for me, and will promote my sanctification, then grant it me, and with it grace to use it to thy glory; but if thou knowest it will prove hurtful, and not conduce to the health of my soul, remove far from me my desire. For every desire that appears to man right and good is not born from heaven; and it is difficult always to determine truly whether desire is prompted by the good Spirit of God, or the evil spirit of the enemy, or thine own selfish spirit; so that many have found themselves involved in evil by the suggestions of Satan or the impulse of self-love who thought themselves under the influence and conduct of the Spirit of God.

Whatever, therefore, presents itself to the mind as good, let it be desired and asked in the fear of God and with profound humility; but especially, with a total resignation of thine own will, refer both the desire itself and the accomplishment of it to Christ, and say, Lord, thou knowest what is best; let this or that be done according to thy will. Give me what thou wilt, and in what measure and at what time thou wilt. Do with me as thou knowest to be best, as most pleaseth thee and will tend most to thy honor. Place me where thou wilt, and freely dispose of me in all things. Lo, I am in thy hands; lead and turn me whithersoever thou pleasest. I am thy servant, prepared for all submission and obedience. I desire not to live to myself, but to thee; O grant it may be truly and worthily. Enable me to die to the honors and pleasures of this fallen world.

It is no small advantage to suppress desire even in inconsiderable gratifications. Self-denial is the basis of spiritual perfection, and he that truly denies himself is arrived at a state of great freedom and safety. See that what is so earnestly sought from God is sought wholly and purely for his honor. That cannot be pure which is mixed with self-interest. Make not, therefore, thine own delight and advantage, but the will and honor of Christ, the ground and measure of all thy requests. For if thou judgest according to truth thou wilt cheerfully submit to his appointment, and always prefer the accomplishment of his will to the gratification of thy desires.

TEMPTATIONS.

Restless and inordinate desires are the ground of every temptation. Many, by endeavoring to fly from temptations, have fallen precipitately into them; for it is not by flight, but by patience and humility, that we must become superior to all our enemies. He who only declines the outward occasion, and strives not to eradicate the inward principle, is so far from conquest that the temptation will recur the sooner and with greater violence, and he will feel the conflict still more severe. It is by gradual advances, rather than impetuous efforts, that victory is obtained; rather by patient suffering that looks up to God for support than by impatient solicitude and rigorous austerity.

That which renders the first assaults of temptation peculiarly severe and dangerous is the instability of our own minds, arising from the want of faith in God. Evil is at first presented to the mind by a single suggestion; the imagination, kindled by the idea, seizes it with strength, and feeds upon it; this produces sensual delight, then the emotions of inordinate desire, and at length the full consent of the will.

It is, indeed, a little matter for a man to be holy and devout when he feels not the pressure of any evil. But if, in the midst of troubles, he maintains his faith, his hope, his resignation, and "in patience possesses his soul," he gives a considerable evidence of a regenerated nature. Some, however, who have been blest with victory in combating temptations of the most rigorous kind are yet suffered to fall even by the lightest that arise in the occurrences of daily life; that, being humbled by the want of power to resist such slight attacks, they may never presume upon their own strength to repel those that are more severe.

Let not strange temptations, that possess thee against thy will, disturb the quiet of thy soul. Maintain only an unchangeable resolution of obedience and an upright intention toward God, and all will be well. It is much safer for most men not to be wholly free from temptation, but rather to be often assaulted, lest they grow secure.

THE WAY TO PEACE.

Behold the way to peace, and to true liberty of spirit. 1. Constantly endeavor to do the will of another rather than thine own. 2. Constantly choose rather to want less than to have more. 3. Constantly choose the lowest place and to be humble to all. 4. Constantly desire and pray that the will of God may be perfectly accomplished in thee and concerning thee. He that doeth this enters into the region of rest and peace.

Let not thy peace depend upon the commendation or censure of ignorant and fallible creatures like thyself, for they can make no alteration in thy real character. True peace and true glory are to be found only in Christ; and he that, seeking them in him, loves not the praise of men, nor fears their blame, shall enjoy peace in great abundance. By love of human praise, and fear of human censure, nothing but disorder and disquietude are produced.

The moment a man gives way to inordinate desire disquietude and torment take possession of his heart. The proud and the covetous are never at rest, but the humble and poor in spirit possess their souls in the plentitude of peace. He that is not perfectly dead to himself is soon tempted and easily subdued, even in the most ordinary occurrences of life. It is not by indulging but by resisting our passions that true peace of heart is to be found. It cannot be the portion of him that is carnal, nor of him that is devoted to a worldly life; it dwells only with the humble and the spiritual.

GENUINE HUMILITY.

Set thyself in the lowest place, and the highest shall be given thee; for the more lofty the building is designed to be, the deeper must the foundations be laid. The greatest saints in the sight of God are the least in their own esteem; and the height of their glory is always in proportion to the depth of their humility. Those that are filled with true and heavenly glory have no place for the desire of that which is earthly and vain; being rooted and established in God, they cannot possibly be lifted up in self-exaltation.

Do not think thou art better than others, lest, in the sight of God, who only knoweth what is in man, thou be found worse. Be not proud of that in which thou art supposed to excel, however honored and esteemed by men; for the judgment of God and the judgment of men are infinitely different, and that displeaseth him which is commonly pleasing to them. Whatever good thou art truly conscious of, think more highly of the good of others, that thou mayest preserve the humility of thy spirit. To place thyself lower than all mankind can do thee no hurt; but much hurt may be done by preferring thyself to a single individual. Perpetual peace dwelleth with the humble; but envy, indignation, and wrath distract the heart of the proud.

The humble man God protects and delivers; the humble he loves and comforts; to the humble he condescends; on the humble he bestows more abundant measures of his grace, and after his humiliation exalts him to glory; to the humble he reveals the mysteries of redemption, and sweetly invites and powerfully draws him to himself. The humble man, though surrounded with the scorn and reproach of the world, is still in peace; for the stability of his peace resteth not upon the world, but upon God. Do not think that thou hast made any progress toward perfection till thou feelest that thou art "less than the least of all" human beings.

To think of having done well without self-esteem is an evidence of true humility, as it is one evidence of great faith to abandon the hope of consolation from created things. Think on the evil that is in thee with deep compunction and self-abhorrence, and think on the good without self-esteem and self-exaltation. There is in thee no good which thou canst glory in as thine own.

The more thou knowest, and the better thou understandest, the more severe will be thy condemnation unless thy life be proportionably more holy. Be not, therefore, exalted for any uncommon skill in any art or science; but let the superior knowledge that is given thee make thee more fearful and more watchful over thyself. If thou supposest that thou knowest many things, consider how many more things there are which thou knowest not at all; and instead of being exalted with a high opinion of thy great knowledge, be rather abased by a humble sense of thy much greater ignorance. And why dost thou prefer thyself to another, since thou mayest find many who are more learned than thou art, and better instructed in the will of God? The highest and most profitable learning is the knowledge and contempt of ourselves; and to have no opinion of our own merit, and always to think well and highly of others, is an evidence of great wisdom and perfection.

SIMPLICITY AND PURITY.

Simplicity and purity are the two wings with which man soars above earth and all temporary nature. Simplicity is in the intention, purity is in the affection; simplicity turns to God, purity enjoys him. No good action will be difficult and painful if thou art free from inordinate affection. And this internal freedom thou wilt enjoy when it is the one simple intention of thy mind to obey the will of God and do good to thy fellow-creatures.

Thy desires must be wholly referred to Christ; and, instead of loving thyself, and following thine own partial views, thou must love only his will, and in resignation and obedience be zealous to fulfill it.

When desire burns in thy heart, and urges thee on some pursuit, suspend its influence for a while and consider whether it is kindled by the love of Christ's honor or thine own personal advantage. If he is the pure principle that gives it birth, thou mayest yield thyself to its impulse without fear; and, whatever he ordains, thou wilt enjoy the event in tranquillity and peace. But if it be self-seeking, hidden under the disguise of zeal for the Lord, this will produce obstruction, disappointment, and distress. It is always necessary to resist the sensual appetite and, by steady opposition, subdue its power; to regard not what the flesh likes or dislikes, but to labor to bring it, whether with or against its will, under subjection to the spirit. And it must be thus opposed, and thus compelled to absolute obedience, till it is ready to obey in all things, and has learned to be content in every condition; to accept of the most ordinary accommodations, and not to murmur at the greatest inconvenience.

THE LOVE OF JESUS.

Blessed is he who knows what it is to love Jesus, and for his sake to despise himself. To preserve this love we must relinquish the love of self and all creatures; for Jesus will be loved alone. If the heart was emptied of self-love and of the love of creatures whom thou lovest only for thine own sake, Jesus would dwell with thee continually. If in all things thou seekest Jesus, thou wilt surely find him in all;

and if thou seekest thyself, thou wilt, indeed, find thyself, but to thine own destruction.

When Jesus is present all is well, and no labor seems difficult; but when he is absent the least adversity is insupportable. When Jesus is silent all comfort withers; but the moment he speaks again the soul rises from her distress. To be without Jesus is to be in the depths of hell; to be with him is to be in paradise. That man only is poor in this world who lives without Jesus; and that man only is rich with whom Jesus delights to dwell. Be humble and peaceful, and Jesus will come to thee; be devout and meek, and he will dwell with thee. Men are to be loved only for the sake of Jesus, but Jesus is to be loved for himself. Jesus alone is to be loved without reserve and without measure; because, of all that we can possibly love, he alone is infinite goodness and faithfulness.

THE PRAISE OF MEN.

He only can have great tranquillity whose happiness depends not on the praise and dispraise of men. If thy conscience was pure thou wouldst be contented in every condition, and undisturbed by the opinions and reports of men concerning thee; for their commendations can add nothing to thy holiness, nor their censures take anything from it. What thou art thou art; nor can the praise of the whole world make thee greater in the sight of God.

The more, therefore, the attention is fixed upon the true state of thy spirit the less wilt thou regard what is said of thee in the world.

If thou hadst but once "known the fellowship of the sufferings of Jesus," and been sensible, though in a small degree, of the divine order of his love, thou wouldst be more indifferent about thine own personal share in the good and evil of the present life; and, far from courting the favor and applause of men, wouldst rather rejoice to meet with their reproach and scorn, for the sake of Jesus. He that loves Jesus, who is the Truth, and is delivered from the slavery of inordinate desire, can always freely turn to God and, raising himself in spirit above himself, enjoy some portion of the blessed repose of heaven.

That man is truly wise, and taught not of men but of God, who perceiveth and judgeth of things as they are in themselves, and not as they are distinguished by names and general estimation. He that has known the power of the spiritual life, and withdrawn his attention from the perishing interests of the world, is not dependent on time or place for the exercise of devotion. He can soon recollect himself, because he is never wholly engaged by sensible objects. His tranquillity is not interrupted by bodily labor or inevitable business, but with calmness he accommodates himself to events as they take place. He is not moved by the capricious humors and perverse behavior of men. If the frame of thy spirit were in right order, and thou wert inwardly pure, all outward things would conduce to thy improvement in holiness, and work together for thy everlasting good. And because thou art disgusted by a thousand objects, and disturbed by a thousand events, it is evident that thou art not yet "crucified to the world," nor the world to thee.

If the truth make thee free, thou shalt be "free indeed," and shalt hear without emotion the commendations or censures of the world. He that liveth not in the presence of Christ, manifested in his heart, is disturbed by the lightest breath of human censure; but he that referreth his cause to the Lord shall be free from the fear of man.

THE CROSS.

In the cross is life, health, protection from every enemy; from the cross are derived heavenly meekness, true fortitude, the joys of the spirit, the conquest of self, the perfection of holiness. Take up thy cross, therefore, and follow Jesus in the path that leads to everlasting peace. The cross is always ready, and waits for thee in every place; run where thou wilt, thou canst not avoid it. And if thou wouldst enjoy peace, and obtain the unfading crown of glory, it is necessary that in every place, and in all events, thou shouldst bear it willingly, and in patience possess thy soul.

The life of Christ was a continual cross, an unbroken chain of sufferings; and desirest thou a perpetuity of repose and joy? To suffer is thy portion, and to suffer patiently and willingly is the great testimony of love and allegiance to thy Lord. It is not in man to love and to bear the cross; to resist the appetites of the body, and to bring them under absolute subjection to the spirit; to shun honors; to receive affronts with meekness; to despise himself, and willingly be despised by others; to bear with calm resignation the loss of fortune, health, and friends; and to have no desire after the riches, the honors, and the pleasures of the world. If thou dependest upon thine own will to do and to suffer all this, thou wilt find thyself as unable to accomplish it as to create another world; turn to the divine power, and the strength of Omnipotence will be imparted.

Thy life must be a continual death to the appetites and passions of fallen nature; and be assured the more perfectly thou diest to thyself the more truly wilt thou live to God. When, therefore, we have read all books and examined all methods to find out the path that will lead us to heaven, this conclusion only will remain, that "through much tribulation" we must enter into the kingdom of God.

Trials will contribute more to the perfection of thy spirit than the gratification of thy will in the enjoyment of perpetual sunshine. The safety and blessedness of man's state in this life are not to be estimated by the number of his consolations, nor by his critical knowledge of Holy Scripture, nor his exaltation to dignity and power; but by his being grounded and established in humility and filled with divine charity, and by seeking in all he doth the glory of God.

LIBERTY OF SPIRIT.

Liberty of spirit cannot possibly be acquired until, with the whole heart, we are resigned, in all situations, to the will of God. Go where thou wilt, rest is not to be found but in humble submission to the divine will. A fond imagination of being easier in any place than that which Providence has assigned us, and a desire of change grounded upon it, are both deceitful and tormenting.

Keep a strict guard over all thy words and actions; let the bent of thy mind be to please Christ only, and to desire and seek after no good but him; and if, with this, thou refrainest from censuring the words and actions of other men, and dost not perplex thy spirit with business that is not committed to thy trust, thou wilt but seldom feel trouble, and never feel it much.

If thy love were pure, and fixed only upon Christ, no creature would have power to enslave thee. Establish thyself in absolute resignation to his good pleasure and thou canst suffer no evil. It is not the acquisition nor the increase of external good that will give thee repose and peace, but rather the contempt of it and rooting the very desire out of thy heart; not only of the luxury of wealth, but of the pomp of glory and the enjoyment of praise. The fruitful root of every evil is thine own unsubdued, selfish will.

Keep invariably to this short but perfect rule: Abandon all and thou shalt possess all; relinquish desire and thou shalt find rest. Revolve this again and again in thy mind; and when thou hast transfused it into thy practice thou wilt understand all things. What can be more at rest than the heart that in singleness and simplicity regardeth only Christ? What more free than the soul that hath no earthly desires?

Nothing should give so much joy to the heart of him that truly loveth thee, O God, and is truly sensible of thy undeserved mercies, as the perfect accomplishment of thy blessed will. He should feel so much complacency and acquiescence as to be abased as willingly as others are exalted; to be as peaceful and contented in the lowest place as others are in the highest, and as gladly to accept of a state of weakness and meanness as others do of the splendid honors and the most extensive power. The accomplishment of thy will and the glory of thy name should transcend all other considerations, and produce more comfort and peace than all the personal benefits which have been or can possibly be conferred.

Often the designs of others will succeed and thine prove abortive; what others say shall be listened to with eager attention, but what thou sayest shall either not be heard or be rejected with disdain; others shall ask once and receive, thou shalt ask often and not obtain; the tongue of fame shall speak long and loud of the accomplishments of others, and be utterly silent of thine; others shall be advanced to stations of wealth and honor while thou art passed by as unworthy of trust or incapable of service. At such trials nature will be greatly offended and grieved, and it will require a severe struggle to repress resentment; yet much benefit will be received from a meek and silent submission; for by such the servant of the Lord proves his fidelity in denying himself and subduing his corrupt appetites and passions.

PATIENCE.

He is not patient who will suffer but a certain degree of evil, and only from particular persons. The truly patient man considers not by whom his trials come, whether by his superior, his equal, or his inferior, whether by the good and holy or the impious and the wicked. But whatever be the adversity that befalls him, however often it is renewed, or by whomsoever it is administered, he receives all with thankfulness, as from the hand of God, and esteems it great gain. There is no suffering, be it ever so small, that is patiently endured for the sake of God which will not be honored with his acceptance and blessing.

Humility and patience under adversity are more acceptable to Christ than joy and fervor when all is prosperous and peaceful. Why art thou offended and grieved at every little injury from men, when, if it were much greater, it ought to be borne without emotion? No evil is permitted to befall thee but what may be made productive of a much greater good. When thou meetest with injury from the violence or treachery of men exert all thy resolution to drive the thoughts of it from thy heart; but if it toucheth thee too sensibly to be soon buried in forgetfulness let it neither depress nor vex thee; and if thou canst not bear it cheerfully, at least bear it patiently.

BRIEF PETITIONS.

Grant, O Lord, that from this hour I may know only that which is worthy to be known; I may love only that which is truly lovely; I may praise only that which chiefly pleaseth thee; I may esteem what thou esteemest, and despise that which is contemptible in thy sight. Suffer me no longer to judge by the imperfect perception of my own senses, or of the senses of men ignorant like myself; but enable me to judge both of visible and invisible things by the Spirit of Truth, and, above all, to know and to obey thy will.

Enable me to die to the riches and honors, the

cares and pleasures of this fallen world; and in imitation of thee, and for thy sake, to love obscurity and to bear contempt. But, transcending all I can desire, grant that I may rest in thee, and in thy peace possess my soul. Thou art its true peace, thou art its only rest; for without thee it is all darkness, disorder, and disquietude. In this peace, O Lord, even in thee, the supreme and everlasting good, I will "sleep and take my rest."

Lord, I will freely suffer for thy sake whatever affliction thou permittest to come upon me; I will indifferently receive from thee sweet and bitter, joy and sorrow, good and evil. For all that befalleth me I will thank the love that prompts the gift, and reverence the hand that confers it. Keep me only from sin, and I will fear neither death nor hell; cast me not off forever, nor blot my name out of the book of life, and no tribulation shall have power to hurt me.

Bring my will, O Lord, into true and unalterable subjection to thine, and do with me what thou pleasest; for whatever is done by thee cannot but be good. If thou pourest thy light upon me, and turnest my night into day, blessed be thy name; and if thou leavest me in darkness, blessed also be thy name; if thou exaltest me with the consolations of thy Spirit, or humblest me under the afflictions of fallen nature, still may thy holy name be forever blessed.

O Lord God, holy Father, be thou blessed now and forever! for whatever thou willest is done, and all that thou willest is good. Let thy servant rejoice not in himself, nor in any other creature, but in thee; for thou only art the object of true joy. Thou, O Lord, art my hope and exaltation, my righteousness and crown of glory.

Grant that I may carefully shun flattery and patiently bear contradiction; that, being neither disturbed by the rude breath of impotent rage nor captivated by the softness of delusive praise, I may securely pass on in the path of life, which, by thy grace, I have begun to tread.

Dearest Jesus, spouse of my soul, supreme source of light and love, and sovereign Lord of universal nature! O that I had the wings of true liberty, that I might take my flight to thee and be at rest! When will it be granted me, in silent and peaceful abstraction from all created being, to "taste and see how good" thou art, O Lord, my God! When shall I be wholly absorbed in thy fullness! When shall I lose, in the love of thee, all perception of myself, and have no sense of any being but thine!

SUGGESTIVE SENTENCES.

Keep thine eye turned inwardly upon thyself, and beware of judging the actions of others. In judging others a man labors to no purpose, commonly errs, and easily sins; but in examining and judging himself he is always wisely and usefully employed.

After all, the most perfect peace to which we can attain in this miserable life consists rather in meek and patient suffering than in an exemption from adversity; and he that has learned most to suffer will certainly possess the greatest share of peace: he is the conqueror of himself, the lord of the world, the friend of Christ, and the heir of heaven!

It requires long and severe conflicts entirely to subdue the earthly and selfish nature and turn all the desire of the soul to God. He that trusts to his own wisdom and strength is easily seduced to seek repose in human consolation; but he that truly loves Christ and depends only upon his redeeming power within him, as the principle of holiness and truth, turns not aside to such vain comforts, but rather exercises self-denial, and for the sake of Christ endures the most painful labors.

All inspection and all judgment being referred to Christ, study thou only to preserve thyself in true peace and leave the restless to be as they will. They cannot deceive Omniscience; and whatever evil they have done or said, it will fall upon their own heads.

Perfection consists in offering up thyself with thy whole heart to the will of God; never seeking thine own will either in small or great respects; but with an equal mind weighing all events in the balance of the sanctuary, and receiving both prosperity and adversity with continual thanksgiving.

With what confidence and peace shall that man, in the hour of his dissolution, look on death whom no personal affection or worldly interest binds down to the present life. When self is once overcome the conquest of every other evil will be easy. This is the true victory, this is the glorious triumph of the new man!

How often has the growth of holiness been checked by its being too hastily made known and too highly commended! And how greatly hath it flourished in that humble state of silence and obscurity so desirable in the present life, which is one scene of temptation, one continual warfare.

The righteous should never be moved by whatever befalls him, knowing that it comes from the hand of God and is to promote the important business of our redemption. Without God nothing is done upon the face of the earth.

"He that followeth me shall not walk in darkness, but shall have the light of life." These are the words of Christ; by which we are taught that it is only by a conformity to his life and spirit that we can be truly enlightened and delivered from all blindness of heart. Let it, therefore, be the principal employment of our minds to meditate on the life of Christ.

A holy life is a continual feast, and a pure conscience the foundation of a firm and immovable confidence in God.

It is an evidence of true wisdom not to be precipitate in our actions, nor inflexible in our opinions; and it is a part of the same wisdom not to give hasty credit to every word that is spoken, nor immediately to communicate to others what we have heard, or even what we believe.

O, if thou didst but consider what peace thou wilt bring to thyself, and what joy thou wilt produce in heaven, by a life conformed to the life of Christ, I think thou wouldst be more watchful and zealous for thy continual advancement toward spiritual perfection.

It is good for a man to meet with contradiction and reproach; to be evil thought of, and evil spoken of, even when his intentions are upright and his actions blameless. For this keeps him humble, and is a powerful antidote to the poison of vainglory.

Spiritual conferences are highly serviceable to spiritual improvement, especially when persons of one heart and one mind associate together in the fear and love of God.

Without love the external work profiteth nothing; but whatever is done from love, however trifling and contemptible in the opinion of men, is wholly fruitful in the acceptance of God, who regardeth more the degree of love with which we act than what or how much we have performed. He doeth much who loveth much; he doeth much who doeth well; and he doeth much and well who constantly preferreth the good of the community to the gratification of his own will.

Endeavor to be always patient of the faults and imperfections of others; for thou hast many faults and imperfections that require a reciprocation of forbearance. If all men were perfect we should meet with nothing in the conduct of others to suffer for the sake of God.

We ought every day to renew our holy resolutions, and excite ourselves to more animated fervor, as if it were the first day of our conversion; and to say, Assist me, O Lord God, in my resolutions to devote myself to thy holy service; and grant that this day I may begin to walk perfectly, because all that I have done hitherto is nothing.

No man can safely go abroad that does not love to stay at home; no man can safely speak that does not willingly hold his tongue; no man can safely govern that would not cheerfully become subject; no man can safely command that has not truly learned to obey; and no man can safely rejoice but he that has the testimony of a good conscience.

Grieve not that thou dost not enjoy the favor of men, but rather grieve that thou hast not walked with that holy vigilance and self-denial which become a true Christian and a devoted servant of God.

While the mind is invigorated by health of body thou wilt be able to do much toward thy purification; but when it is oppressed and debilitated by sickness I know not what thou canst do. Few spirits are made better by the pain and languor of sickness.

It is better to turn away from all that produces perplexity and disturbance, and to leave everyone in the enjoyment of his own opinion, than to be held in subjection by contentious arguments.

All is vanity but the love of God and a life devoted to his will.

"CHRISTIAN PERFECTION."

Alphonsus Rodriguez, author of one of the very best works on Christian Perfection which has ever seen the light, was born at Valladolid, in Spain, in 1526. He received the degree of Doctor in Philosophy at the University of Salamanca, and afterward, discharging the office of professor of moral philosophy, became so famous for his lectures that students flocked to hear him from all parts of the country. After twelve years of this public teaching he devoted himself for the remainder of his life to imparting spiritual instruction to young priests and monks, and he soon came to be looked upon as "one of the greatest masters of the science of the saints and the conduct of souls." In Cordova, Seville, Valladolid, and Montilla he spent his time doing good until his departure to a better world, in 1616, having been greatly honored and loved by all who knew him. It is written of him that "he lived so entirely detached from himself, and from every feeling of self-love, that he regarded God alone in all things. He showed an ardent zeal for the salvation of souls, and left the world an heroic example of holiness. Whatever leisure time he could spare from his indispensable occupations he employed in mental prayer and in spiritual reading. He taught nothing which he did not himself practice, and his book is but the mirror of his life."

The book, *The Practice of Christian Perfection,* is the mature result of his lifelong study, and a digest of the directions which he was accustomed to give to those under his care. It was first published at Seville in 1614, and was soon translated into all the languages of Europe. Of French translations there are no fewer than six. The work is divided into eighteen parts, and each part is subdivided into many chapters. The unabridged edition is in three large volumes, but a revised edition, more suitable to the majority of Christians, is published in two small volumes by Burns & Oates, of London. Most of it is as well adapted to the profit of Protestants as of Roman Catholics; but the selections which we append, since they contain the cream of the distinctive teachings of the author, will doubtless be found sufficient for ordinary readers.

SPIRITUAL ADVANCEMENT.

One of the principal causes of the little progress we make in holiness is that we do not desire and long for it with sufficient earnestness; we desire it, it is true, but so feebly and coldly that the desires we form vanish almost as soon as they are conceived.

It is said of Apelles that, in whatsoever business he was engaged, he never let pass a day without exercising himself in his own profession by painting something or other. For this purpose he always endeavored to find out some time amidst his other employments, and to excuse himself from going into company was wont to say, "This day I have not as yet drawn one stroke with my pencil;" so that by this means he became a most excellent painter. In like manner you will become an excellent Christian if you let no day pass without making some advancement in virtue. Practice daily some act of mortification, correct some fault you were accustomed to commit, and you will quickly find that your life will become every day more perfect. When you examine your conscience at noon and perceive that you have done nothing that morning conducive to your improvement, that you have mortified yourself in nothing, that you have performed no act of humility when occasions offered themselves, believe that you have lost so much time, and make a firm resolution not to let the remaining part of the day pass in the same manner. You will find it impossible to observe this rule without gradually advancing, and making by degrees a considerable progress in the way of perfection.

"The path of the just is as the shining light, that shineth more and more unto the perfect day." The just man never believes that he has fully performed his duty; he never says it is enough, but always hungers and thirsts after righteousness; so that if he were to live here forever he would perpetually strive to become more righteous and more perfect, and to advance always from good to better.

We must never think we are holy enough, but always aspire to become still more so. Whoever wishes to be a saint must forget what he has done and constantly think on what he has still to do. He is truly happy who advances daily, and who never thinks on what he did yesterday but what he has to do to-day in order to make new progress. The former tempts us to repose, the latter incites us to go on. It is a great shame and confusion to us that worldly men desire those things that are pernicious to them with more earnestness than we desire those things that are of the greatest advantage, and that they run faster to death than we do to life.

BROTHERLY LOVE.

The love which each of us owes God is a debt he has transferred to our neighbor; and the charity you exercise toward your brother you exercise toward God, who receives it as if it were done to himself. This ought to be a powerful motive to excite us to love our brethren, and do them all the good we can; because though it seems to us that we do it to those to whom we owe nothing, yet if we look upon God, and reflect upon the infinite obligations we are under to him, and consider that he has transferred all his right to them, we shall find that we are indebted to them for all we have.

One of the things by which we ought most of all to testify the love we have for our brethren is the speaking of them in such manner as to make known to others the esteem we ourselves have for them. Though your brother has his defects, it is hard also if he should not have something commendable in him. Imitate the bee, which lights upon flowers only, not minding the thorns that surround them; and follow not the example of the beetle, which lights upon nothing else but dirt.

Never speak ill of your neighbor or discover his defects, though ever so small or apparent. Never do him any prejudice, or let the least contempt of him appear, either in his presence or absence. Never tell anyone what has been said of him when the thing may give him the least offense, because this is to sow discord amongst brethren.

Never break out into passionate words, nor say anything mortifying to your neighbor, nor be obstinate in your own opinion, nor dispute nor contest with heat, nor reprehend anyone over whom you have no authority. Behave sweetly and charitably to everyone, doing everything in your power to serve others and make them happy. And if by your office you are in a more special manner bound to serve your neighbor and to take care of him, you must apply yourself to it still more particularly, and endeavor, by sweetness of manners, of words, and answers, to supply whatever is not in your power to do for him.

Never judge your neighbor, but endeavor to excuse the faults he commits against others and against yourself; and in general have a good opinion of everyone, harboring no aversion nor showing the least sign thereof, either by abstaining from speaking to him or neglecting to succor him in his necessities.

PRESENCE OF GOD.

To employ ourselves continually in the exercise of the presence of God is to begin in this life to enjoy the felicity of the blessed in the next. The saints and patriarchs of the Old Testament took particular care to walk always in God's presence. Without doubt they impose upon themselves a strict obligation to live well who consider that all they do is done in the presence of a Judge who sees all, and from whom nothing can be concealed. If the presence of a grave person is sufficient to keep us to our duty, what effect ought not the presence of the infinite majesty of God to produce in us? What servant is there so insolent as to despise his master's orders in his very presence?

The presence of God is a sovereign and universal remedy for all the temptations of the devil and all the repugnances of nature. So that if you desire a short and easy means to gain perfection, and such a one as contains within itself the force and efficacy of all others, make use of this which God himself gave to Abraham, "Walk before me, and be thou perfect."

See how the moon depends upon the sun; see how necessary it is for her to keep her face always to it. As soon as anything interposes between the sun and moon the moon presently loses its light and force. The same thing happens between the soul and God, who is its sun; and it is for this reason that the saints so earnestly exhort us to have the presence of God constantly before our eyes.

To place ourselves in the presence of God it is not necessary we should represent him as by our side, or in this or that particular place, nor imagine him as under such or such a form. What we are to do is to believe, as a certain truth, that he is really and effectually everywhere. But we must not only employ our understanding to consider God as present; we must afterward exercise our will in loving him, and in uniting ourselves to him as present; and it is in this that the chief exercise of the presence of God consists.

The act of the will by which we must elevate our hearts to God in this exercise consists in the ardent desires of the soul to unite itself to him in the bond of a perfect love. These desires and aspirations are expressed by short and frequent prayers, which are called "ejaculatory," that is to say, "suddenly shot forth," because they are like inflamed darts or arrows which the heart shoots, one after another, toward God.

St. Basil makes the practice of this exercise to

consist in taking occasion from all things to call God to mind. If we eat, let us give thanks to God; if we clothe or dress ourselves, let us always render him thanks; if we look up to the heavens, let us praise God who created them; and as often as we awake in the night, let us never fail to elevate our hearts to God. Endeavor in all things you do to elevate your heart to him, saying, "Lord, it is for thy sake I do this; it is to please thee; all my joy, all my satisfaction is the fulfilling of thy will, and so that I do but please thee I desire nothing more." This is a most excellent and perfect way of walking in God's presence; because it is to entertain ourselves in a continual exercise of the love of God. Of all the means we can imagine there is no one better or more profitable than this, to keep ourselves always in that continual prayer which our Saviour requires we should practice.

Moreover, we must take notice that when we make these acts, and say these petitions, we must say them, not as elevating our heart or raising our thought to something without us, but as speaking to God present within us; for this is properly to walk in the presence of God, and this is what will render this more sweet, pleasant, easy, and profitable to us than any other sort of prayer whatsoever.

But that which we must most of all take notice of is that when we put ourselves in the presence of God it is not to remain or rest there, but that this presence may serve as a means or help to perform all our actions. For if we content ourselves with barely attending to the presence of God, and so become negligent in our actions, this attention would be no profitable devotion, but a very hurtful illusion. Whilst, therefore, we have one eye engaged in contemplating God, we must engage the other in seeing how to do all things well for his love; so that the consideration of our being in his presence may be a means to oblige us to do all our actions better.

CONFORMITY TO THE WILL OF GOD.

Our perfection consists in conformity to the will of God, and the greater this conformity is the greater also will be our perfection. Perfection essentially consists in the love of God, and the more we love God the more perfect we shall be. But as the love of God is the most elevated and most perfect of all virtues, so the most sublime, the most pure, and the most excellent practice of this love is an absolute conformity to the divine will. Moreover, it is certain that there is nothing better or more perfect than the will of God, and consequently we shall become better and more perfect in proportion to our union with this will.

There can nothing happen in this world but by the order and will of God. And this is always to be understood except of sin, of which he is neither the cause nor author. Sin excepted, all other things, as sufferings, pains, and afflictions, happen by the order and by the will of God. This is a truth not to be called in question; for, though all these things proceed from second causes, it is certain that there is nothing done throughout the universe but by the command and will of one sovereign Master who orders and governs all. There is nothing that happens by chance. There is not a leaf that stirs upon a tree but by his will. And it is by this will that those things are regulated in which chance seems to have a greater share.

We ought to infer from these truths that we must receive all things as coming from the hand of God, and in them conform ourselves entirely to his divine will. We must look upon nothing as happening by chance, or by the conduct or malice of man; for this is what ordinarily is wont to give us most trouble and pain; nor must we imagine that this or that thing has happened to us because such or such a one had a hand in it; nor that, if such or such an accident had happened, things would have fallen out after a different manner. About this we must not amuse or trouble ourselves; but in what way or what manner soever anything happens to us we must always receive as coming from the hand of God, because it is he in reality who by these means sends it to us. An ancient father in the desert was wont to say that a man would never enjoy true peace and satisfaction in this life till he could persuade himself that only God and he were in this world.

It is a truth so firmly supported by the authority of Holy Scripture that all misfortunes and sufferings come from the hand of God, that it would not be necessary to prove it at greater length if the devil, by his vain subtleties, did not endeavor to obscure it and render it doubtful by insisting that the evils which happen by means of man proceed only from malice and sinful will. When we have anything said against us we imitate dogs which, when a stone is thrown at them, run to bite it, and take no notice of him that threw it; so we take no notice of God who sends us these mortifications, but run after the stone and make an attack upon our neighbor.

Observe that in every sin we commit there are two things. The one is the motion or exterior act, the other the irregularity of the will, by which we transgress what the commandments of God prescribe. God is the cause and author of the first; man only is the cause and author of the second. It is God who produces the motions, as he produces all other effects that proceed from irrational creatures. For, as they cannot move themselves, or act without God, so neither can man without his help move his arms or other limbs. Moreover, these kinds of natural motions or actions have nothing in them that is bad; because, if a man should make use of them, and either for his own defense, or in a just war, or as a minister of justice, should kill another, it is certain he would not commit any sin. But in what makes the action sinful—that is to say, the irregularity of the will that moves or determines him to commit a murder—he is not the cause of it. The truth of this is explained by the following comparison: One has received a hurt in his foot which makes him lame. What causes him to walk is the faculty and power he has to move himself, but what causes him to halt is the hurt in his foot. It is the same in every vicious or sinful action. The cause of the action is God; but the cause of the sin, mixed with the action, proceeds from the free will of man.

So that God neither is nor can be the cause or author of sin. But as to other evils, whether they proceed from natural causes and irrational creatures, or whether they come from men, or from whatever other source they spring, or in whatever manner they may happen, we must believe for certain that they proceed from the hand of God, and happen to us by his divine providence. It is God that moved the hand of him that struck you, it is God that gave motion to his tongue who gave you injurious language. "Shall there be evil in a city," says the prophet Amos, "and the Lord hath not done it?" Which truth the Holy Scripture frequently takes note of, often attributing to God the evil which one man does to another, and saying that it is God himself that has done it.

Those who have attained a perfect conformity to the divine will, and who place their own contentment in that of God, never suffer themselves to be disquieted at the changes and accidents of this life. Their will is so fully subjected to that of God that the very assurance they have that all things come as sent by him, and that his holy will is accomplished in whatever adversity happens to them, makes them, by preferring his will to their own, look upon all their tribulations and sufferings as so many joys, and all their griefs and sorrows as so much sweetness and consolation. Hence it is that nothing can trouble them; for as trouble can come only from crosses, misfortunes, or affronts, and as these, through respect for the hand which sends them, are received by them as so many favors, it follows that there is nothing which can change or diminish the peace and tranquillity of their soul. Each day of their life is a day of jubilee and exultation. Having attained a perfect conformity to the divine will, they meet everywhere sources of content and satisfaction.

The holy abbot Deicola is said always to have had a smile on his countenance; and, being once asked why he was uniformly so cheerful, he answered that it was because no one could deprive him of Jesus Christ. He had experienced a real content since he had placed all his felicity in that which could never fail and which could never be taken from him.

It is certain if you never desire anything but what God desires you will always attain the object of your desires, because God's holy will can never fail of being entirely performed. How happy we when we covet nothing but what God pleases! And how happy, not only because our own will is accomplished, but because we see the will of God, whom we love, accomplished in us and in all things. It is the second consideration on which we ought chiefly to dwell; and it is only in the contentment of God, and in the execution of his holy will, that we ought to place all our joy and satisfaction.

This conformity to the will of God is a most efficacious means for the attaining of all other virtues. Exercise in the one is exercise in all. Occasions occur every moment of practicing humility, obedience, patience, and the rest. The obtaining the one virtue will put us in possession of all. If you desire an easy and compendious way of attaining perfection, here you have it. Say daily, "Lord, what wouldst thou have me to do?" Have always these words in your mouth and heart; and according as you strengthen yourself in these holy sentiments, so will you increase in the perfection you aim at.

We must endeavor, in our prayers, to reduce by frequent acts this exercise to practice; and never cease searching this rich vein of God's fatherly providence over us till we have found the inestimable treasure of a perfect conformity to his holy will. I am certain, let us say, that nothing can happen to me without his orders, and that neither men, nor devils, nor any other creature whatever can effect anything contrary to his holy pleasure. I will, then, refuse nothing he sends, and I will desire nothing but the accomplishment of his will.

Let us, then, make it our endeavor to become such by God's holy grace that we may receive with joy and satisfaction whatever misfortune happens; and find so great a satisfaction in whatever proceeds from the divine will as thereby to sweeten all the bitterness of this life, and make whatever is hard and difficult easy and delightful. We ought to shut ourselves up in the divine will as in a most secure retreat, and live there as a pearl in the shell, or as a bee in the hive, without ever coming forth. At first we may find the place very narrow, but afterward it will be larger; and without once coming forth we may walk there as in the habitations of the blessed.

MEDITATION.

Meditation is the beginning and ground of all good. It is the sister of spiritual reading, the nurse of prayer, and the director of good actions. It causes true devotion to spring up in our hearts. It is that which, next to the grace of God, most of all warms the heart and the will, and produces the prompt disposition to do virtuous deeds. So that true devotion and fervor of spirit consist not in a certain sensible sweetness which some experience in prayer, but in having our will always disposed and ready to execute what may in any way conduce to God's glory and service. Since we make use of meditation and reflection to excite our will to act, and since this is our only aim and end, we must not entertain ourselves in meditation any longer than is necessary to move our will.

No one becomes perfect on a sudden; it is by mounting, and not by flying, that we come to the top of the ladder. Let us, therefore, ascend, and let meditation and prayer be the two feet we make use of to do so. For meditation lets us see our wants, and prayer obtains for us relief from God. The one makes us discern the dangers that surround us; the other gives us happy escape from them. Prayer is tepid without meditation.

OBSTINACY.

Obstinacy, though it be in a matter of truth, can come from none but the devil. The reason is, because that which usually moves a man to maintain his own opinion with any heat is the desire he has of being esteemed. Hence it happens that, to appear more able or learned than his adversary, he endeavors to convince him that he is in an error; and if he cannot be victorious in his dispute he endeavors at least to make it appear he had not the worst of it; and thus it is always the demon of pride who is the occasion of this obstinacy.

The spirit of contradiction is a very bad one; endeavor, therefore, to cast it out, though the thing in question be of consequence. If anyone should contradict you insist not much upon it, nor suffer yourself to be carried on by a desire of getting the better of him; but explain yourself once or twice with all possible mildness, and show him your idea of the question, and after that let him believe what he pleases; and impose silence upon yourself as if you had nothing more to say about the matter.

It is related of St. Thomas Aquinas that in his disputations he always proposed his opinion with meekness and sweetness, with an unspeakable moderation, without any show of presumption, and without the least offense to anyone; but behaved as a man who regarded not gaining the victory, but merely endeavored to make known the truth.

We read of Socrates that, dining one day with his friends, and happening in a large company to rebuke a little too sharply one of the guests, Plato, who was present, could not refrain from saying to him, "Would it not have been better for you to have deferred this rebuke to another time, and secretly to have told him of his fault?" "But would not you also," replied Socrates, "have done much better to have told me of mine in private?"

JOY AND SADNESS.

Sadness is a disease more dangerous and difficult to be cured than all other spiritual infirmities. Beware of admitting it into your soul, for if it once gets possession of you it will soon take away all your relish for prayer and spiritual reading. Sadness makes us severe and rude to our brethren. It renders a man impatient, suspicious, and intractable; and sometimes it so troubles our mind that it even deprives us of our judgment. Sadness in the heart of a Christian is a subject of joy to the devil, because then it is easy to make him either despair or turn to the pleasures of the world.

God desires to be served with joy. When we serve him thus we promote his honor and glory, because we show that we do it with affection, and that all we do is nothing compared to what we would wish to do. God is not only more honored in this way, but our neighbor also is more edified, and the esteem of virtue more increased; for those who serve God with joy prove to worldlings that on the road of virtue there are not so many obstacles and difficulties as is imagined; and as men naturally love joy they willingly travel the road in which they expect to find it.

The saints look upon cheerfulness as so great a good that they say we ought not to be discouraged or made sad even in our spiritual falls. Our sadness should at least be moderated by our hope of pardon and our confidence in God's mercy. Fathers behold the falls of their children rather with compassion than anger; God does the same to us.

There is a sadness which is according to God, and one which is not. The first is obedient, affable, humble, sweet, and patient, and since it proceeds from the love of God it preserves in us the fruit of the Holy Spirit. The other sorrow is rude, impatient, and full of disquiet and bitterness; it hinders us from what is good, and produces discouragement and despair.

The sadness that is holy proceeds from a sight of our sins, or from a consideration of the many sins daily committed in the world, or from a great desire of perfection and the little progress we make toward it, or from a sacred impatience of visiting our celestial country.

The joy of the servants of God is not a vain and frivolous one; it is not a joy that makes us break out into loud laughter, or to say witty things, or to join in conversation with everyone we meet. For this would be a dissipation of mind, immodesty, and irregularity. The joy we seek is a prudent one, that comes from within and is visible in our countenance without. We read of many saints who had such a joy and serenity in their looks that it gave testimony of the peace and satisfaction which they enjoyed in their hearts. And this is the joy which we should all possess.

TEMPTATIONS.

To encourage us in our temptations it will be a very great help if we consider the weakness of our enemy, and how little he is able to do against us; seeing that he cannot make us fall into any sin against our own will.

Prayer is one of the principal means by which we resist temptation. As a man who lies at the foot of a tree and sees wild beasts coming toward him to devour him would presently climb to the top of it to save himself, so one who perceives himself beset with temptations ought to climb up to heaven and retire into the bosom of God by means of prayer, and thus he will be delivered.

The general maxim to defend ourselves from any temptation is presently to have recourse to what is most contrary to it. We must cure those temptations we are most subject unto by practicing what is contrary to them. For example, when we find ourselves carried away with vanity and pride we ought to exercise ourselves in servile works, and so on all other occasions steadfastly resist our bad inclinations.

Another excellent remedy is strongly to resist temptations in their beginning. Another is to be always employed. Do not dwell upon your temptations. They are like little dogs that bark after a man that passes by; if he stops to drive them away they bark more fiercely than they did before. We must therefore do like him who walks in a street where the wind blows the dust in his face; he covers his eyes and walks on his way without troubling himself either with the wind or the dust. When any bad thought occurs we must endeavor to turn our mind from it by applying it to something else; for example, by thinking on the death and passion of our Saviour, or some such object. However wicked and shameful the thoughts may be that arise within us, if, instead of entertaining them, we are troubled at having them, so far from believing that God has forsaken us we should consider it an infallible sign that he remains within us, because it is he alone who is able to give us this horror of sin and this fear of losing his grace.

PERFECTION OF OUR ORDINARY ACTIONS.

It is in performing well the most common and familiar actions of our life that our advancement and perfection consist. We shall become perfect if we perform these perfectly; we shall be imperfect if we perform them imperfectly. And this is all that properly makes the difference between a perfect and an imperfect Christian. For our perfection arises not from our doing more things than another does, but from our doing them better; and in proportion to the manner in which a man does these works will he become more or less perfect.

The goodness of our actions consists of two things, of which the first and chief is, that we act purely for God. The intention is the foundation of the goodness of all our actions. The second is, always to walk in God's presence. Thus shall we be always in prayer. They pray always who always perform their actions to please and glorify God; thereby they make their life a perpetual prayer.

The third means of doing our actions well is, to do each one as if it were the only one we had to do. Another means is, to do each action as if it were to be the last we were to perform in this life. One of the best means to know certainly whether we walk upright before God is, to consider whether we are in a state to answer him at whatever time he calls, and in whatever employment we are engaged.

Perform with exactness what belongs to your office and employment, and use all possible care and application in it, as doing all things for God and in his presence. Do not commit deliberately any fault, however small. Set great value on even the least things. And since our own spiritual advancement depends upon the due performance of our ordinary actions we must, from time to time, as soon as we perceive we begin to relax in any one, take care to make it the subject of our particular examination; and so renew by this means our fervor and attention.

MORTIFICATION, OR SELF-DENIAL.

Begin this exercise in profiting by those occasions of mortification which are daily offered you by your superiors, by your brethren, or in any other way. Receive all with a good will, and make your profit of them; seeing that they are the things necessary for your own peace, as well as for the edification of your neighbor. If we will profit by all the occasions of mortification that happen to us from our neighbors or brethren we shall meet with a sufficient number, and of all kinds. Some will mortify us intentionally, others through negligence but without a bad intention, while others will mortify us either from contempt or a want of due esteem for us. But if we consider those which God sends us directly, as sickness, temptations, disquiet of mind, the unequal distribution of his gifts, as well natural as supernatural, we shall find them to be numberless.

As those who design to make themselves soldiers practice in time of peace military evolutions, which, though but mock fights, yet qualify them for real combats, even so the Christian must endeavor to mortify himself and renounce his will in small things which are lawful for him to do, that he may be the more ready to mortify himself in those which are forbidden. If we accustom ourselves to renounce our will in these small things, and things that are indifferent, we shall the sooner be able to deprive ourselves in greater.

Take care to do nothing, to think nothing, to speak nothing, purely to please your own will, or to satisfy your sensual appetite. Before meals mortify in yourself the desire of eating; and eat not to satisfy your appetite, but to obey God, who will have you eat to nourish yourself. Before you go to study mortify your desire, and then study because God commands you to do so, and not because you find pleasure in it. Before you go into the pulpit to preach, or to explain any public lesson, mortify your own desire of doing it, and then preach or teach not because you like to do it, but because it is what you are commanded to do, and because it is God's will. Observe the same practice in all other things; and thus, depriving your actions of the attachments you have to them, perform them all purely for God's sake.

To accustom ourselves, in all our actions, not to do our own will, but God's, and to take delight in them not because they are pleasant in themselves and because our inclination moves us to perform them, but because in doing them we do the will of God—this is a point of great importance, and having in it a high degree of spirituality. He who performs his actions in this manner will at the same time accustom himself not only to mortify his own will, but also to do the will of God in all things, which is an exercise of the love of God most profitable and most perfect. We should at all times entertain a holy joy that the will of God is fulfilled in us.

The progress of a Christian consists not in a happy disposition, in an agreeable exterior, in a sweet temper, but in our endeavors to overcome ourselves and in the victory we gain over our passions. This in an infallible test of anyone's advancement in perfection; and therefore one who is naturally choleric does far more, and merits a greater recompense, when he overcomes his passion, than you who naturally are of a milder disposition, and who have nothing to resist or overcome. Neither the sweetness of your temper nor the natural heat or impetuosity of another ought to make you esteem yourself the more or him less. On the contrary, you must make it an occasion of humbling yourself, acknowledging that what appears to be virtue in you is not so, but an effect of your natural temper, and that it is a great virtue in others to do the same things you perform.

There are three degrees of mortification which are steps to raise us to the highest pitch of perfection. The first is taught us by St. Peter: "Dearly beloved, I beseech you as strangers and pilgrims, abstain from fleshly lusts, which war against the soul" (1 Pet. ii, 2). The second is far more sublime than the first, and is thus described by St. Paul: "Ye are dead, and your life is hid with Christ in God" (Col. iii, 3). A man that is dead is equally indifferent to praise and censure, is unconcerned at any contempt or injury done him, no passion of pride or anger disturbs him, nothing troubles him. If, then, you still have eyes to pry into other people's actions; if you are never at a loss for an answer to excuse yourself, and to do away the obligation of obedience; if you take it ill when you are reproved; and, lastly, if you feel proud and angry when you are neglected or despised, be assured that you are so far from being dead to the world that you live and act by a worldly spirit. But there is a third degree of mortification. To die upon a cross is more than barely to die; it is to die a death of the greatest infamy. To this degree St. Paul was raised when he said (Gal. vi, 14): "The world is crucified unto me, and I unto the world." It is the same as if he had said, Pleasures, honors, riches, the esteem and praise of men, and all that the world courts and adores, is a sensible cross to me; on the other hand, I love and embrace with the greatest delight all that the world looks upon as infamy and disgrace. To be insensible to affronts and disgrace is a small matter in his sight who rejoices and glories in them, and says with St. Paul, "God forbid that I should glory, save in the cross of our Lord Jesus Christ, by whom the world is crucified unto me, and I unto the world."

RASH JUDGMENTS.

The first root whence rash judgments commonly grow is pride, which, though it is the root of all other sins, yet is much more particularly so of this. Those who think themselves somewhat advanced in a spiritual life are more frequently tempted than others to judge and censure their neighbors, forgetting their own defects.

The saints say that simplicity is the daughter of humility; for he who is truly humble has not his eyes open to see the faults of his neighbor, but only to discern his own; and finds so many things to consider and deplore in himself that he never casts his eyes or thoughts on the failings of others. If, therefore, we were truly humble we should be far from these kinds of judgments. The sight of our own defects gives us humility and contrition, augments the fear of God in the soul, keeps it in recollection, and produces in it the fruits of peace and tranquillity. On the contrary, the practice of observing the faults of others is the cause of many evils and inconveniences; it carries along with it pride, rash judgments, indignation against our neighbor, contempt of our brethren, remorse of conscience, indiscreet zeal, and a thousand other imperfections which agitate and injure the heart.

Though there is no sin in judging that an action is bad when it is evidently so, yet should that which we see be manifestly culpable it is still a virtue to endeavor, as far as in our power, to excuse our brother. Excuse the intention if you cannot excuse the action; believe it proceeds from ignorance or surprise; that it is an effect of the first motion which he was not master of. If we loved our brethren as ourselves we should not want reasons to excuse them. Self-love always furnishes us with an infinity of excuses; it puts arms in our hands to defend ourselves and teaches us how to lessen our own faults; and without doubt we should make use of the same means in behalf of our neighbor if we loved him as we love ourselves.

When we have a passionate affection for anyone we approve of all his actions, and are so far from giving them any bad interpretation or taking them in ill part that, though we cannot but see his faults, yet we think of nothing else than how to palliate and diminish them as much as we are able. The same fault, accompanied with the same circumstances and appearances, seems not to be the same in him we love as it does in him we have no affection for.

VAINGLORY.

Vainglory is a sweet-scented powder, but it is entirely composed of arsenic. It corrupts and destroys all the merits of our actions after they are done, and makes us lose all the advantages we might expect from them. It waits till we have taken pains to perform many good works, and afterward it robs us of them. It is like a pirate that attacks not a vessel which is sailing out of port to purchase goods, but waits till it returns home richly freighted, and then fails not to attack it. Vainglory turns good into bad, virtue into vice, through the vanity of the miserable end we purpose to ourselves; hence, instead of the recompense due to us, it causes us to merit nothing but punishments. It is a tempest in the harbor. It does to the most perfect Christian what a man does who, going on board a ship well provisioned and richly laden with merchandise, bores a hole in the bottom of it through which the water enters and at length sinks it.

The first remedy against vainglory is to consider with attention that the good opinion of men is but mere wind and smoke, because it neither gives nor takes anything from us, whether good or bad; neither makes us better nor worse. A second remedy is to take very great care never to use any expression in praise of ourselves. Never say anything of yourself that may redound to your praise, though the person you speak to should be one of your most familiar friends. If it seems necessary for the instruction of others to say something of edification that has happened to yourself relate it as of a third person.

We must go yet further, and even conceal as much as we can the good actions we perform. It is after this manner that travelers hide their money with a great deal of care, lest they should be robbed of it. Some have compared those who perform their good works through a spirit of pride to hens who make a cackling after they have laid an egg, whereby they cause it to be discovered and lose it in consequence. The true servant of God esteems the good he does as nothing; and what he cannot hide from the eyes of men he believes he has already received a kind of reward for, if he adds not other good works which cannot come to their knowledge. Do not, therefore, aspire to the esteem of men, for fear that God should make that to be the extent of all the recompense of those good actions you were able to perform.

It is for the same reason the saints counsel us to avoid all sorts of singularity in devotion, because singular and unusual actions are most remarked and most spoken of. And he who does what others do not draws the eyes of all the world upon him; whence arises the spirit of pride and vainglory which makes us look upon others with contempt.

But because we cannot always hide our good actions, since some are obliged to contribute by their example to the edification of their neighbor, the first means of defense against vainglory is to rectify in the beginning our intention, and to elevate our heart to God, and offer him all our thoughts, words, and actions, to the end that when vainglory comes to claim a part in them we may say to it, You come too late; all is already given to God.

We read of a father in the desert who used to pause a little before performing any action. One day being asked why he did so, "I believe," said he, "that all our actions have no merit of themselves, if they be not done for a good end. Wherefore, as he who fires at a target takes his aim for some time in order to cover the object, even so, before I perform what I purpose, I direct my intention to God, who ought to be the only object or end of all our actions; and it is upon this account that I always pause a little at the commencement of every action."

PRAYER.

We must not confine ourselves to prayer as the end in which we are to repose; it is only the means we make use of to advance ourselves in perfection.

Let each one consider for some time before he begins his prayer, and let him ask himself: "What is the greatest spiritual infirmity I have? What is the obstacle that most opposes itself to my progress in virtue?" Do not go to prayer, like a hunter that shoots at random, with a vague design of profiting by what may be presented to your mind. Take to heart for some time some one thing in particular; that which you find yourself most in need of. We must chiefly insist upon this and beg it of God with fervor several times, several days, nay, even several months, making this our chief business, having it continually before our eyes, till we come at last to obtain it. "One thing I have asked of the Lord," says the psalmist; "this will I seek after." It is of great importance to dwell upon one thing till the soul is well filled and penetrated with it.

It was the practice of a very great saint that when her heart was silent she neglected not to speak with her lips, because she thus renewed and enlivened the fervor of her heart, and she also confessed that sometimes, for want of making vocal prayer, when she found herself sleepy, she also omitted her mental prayer. This is but too often experienced; tepidity and drowsiness to which we give way in time of prayer are the causes why our lips are silent; but if we forced ourselves to speak we should overcome these impediments, and should animate ourselves with new fervor.

Our praying well, and consequently our acting well, during the whole day depends much on our seizing the first moments of the morning, as soon as we wake, to preoccupy them with good thoughts. We must be extremely vigilant, in order that, as soon as our eyes are open, our imagination may be filled with the thought of God, and our memory and heart receive a similar impression before any strange thought is able to make its entrance.

Another profitable advice is to write down very briefly the fruit we have reaped from prayer; the good thoughts we have had, the pious resolutions we have made, and the lights we have received from God in it. By this means the good desires and resolutions we make are more perfected and take deeper root, and make a stronger impression on our heart; and experience also will teach us that when at another time we come to read them over again they will be of great profit to us.

HOW TO BE HUMBLE.

Humility is the source, foundation, and root of all virtues, as pride is the beginning of all sin. All virtues which are not founded upon humility are virtues only in appearance.

Root out of your heart pride, and plant humility in its place. As soon as you shall be truly humble you will be obedient, you will be patient, you will complain of nothing, you will think nothing hard; and though anything should happen to you very difficult to be borne with, yet it will always seem to you very little in comparison with what you deserve. As soon also as you shall be humble you will be charitable toward your brethren, because you will believe them all to be good, and better than yourself; you will have a great simplicity of heart, and you will judge ill of nobody, because you will have so great a sorrow and confusion for your own defects that you will not think at all upon those of your neighbor. The love of God is very much increased by means of humility; for one of a humble spirit, seeing that he receives whatever he has from the hand of God, and that he is very far from meriting it, feels himself excited to love his benefactor more and more. The humble man is not angry at others being preferred to him; he is willing that they should be esteemed and himself despised. There is no envy among the humble. If you seek a ready way to acquire all virtues, and a short lesson for attaining perfection, you have it in two words: *be humble.*

Humility consists not in words, nor in outward conduct, but in the sentiments of the heart; in having a low and mean opinion of ourselves, founded on the deep sense we have of our own nothingness. It is a virtue by which a man, from a true knowledge of himself, becomes vile in his own eyes. It is needful that before all things you should know yourself thoroughly, and after that esteem yourself according to what you are. You will be humble enough as soon as you know yourself; for then you will plainly see how little you are. According to some, one of the reasons why God loves humility is because he loves the truth above all things. Humility is truth itself, whereas pride is a mere deceit and a lie; for you are not in reality what you think you are, nor what you would have others think you to be.

But, lest we be overmuch cast down at the sight of our imperfections, we should, for our encouragement, immediately pass on to the consideration of God's goodness. Yet there is danger in dwelling too much upon this latter. Our exercise ought to be like Jacob's ladder, of which one end touched the earth and the other reached up to heaven. It is by it you are to ascend and descend, as the angels did. Ascend till you arrive at the knowledge of the goodness of God; but rest not there, for fear of falling into presumption. Go down again forthwith to the knowledge of thyself; and rest not there either, for fear of being faint-hearted, but return up again to the knowledge of God, to place all your confidence in him. In fine, all you have to do is to go continually up and down this ladder. These are the two lessons which God gives every day to his elect—one to consider their own faults, and the other to consider the goodness of God, who, with so much bounty and affection, pardons them.

Gerson makes an ingenious application of the fable of Antæus to the subject we speak of. The poets feign that Antæus was a giant and son of the earth, who, having been thrice thrown to the ground while he wrestled with Hercules, regained additional strength every time he touched the earth. Hercules, perceiving this, raised him up from thence, and squeezed him to death in his arms. This is a figure of what the devil does when he fights with us; he endeavors to lift us up very high by means of the esteem and praise of men, that so he may the more easily overcome us. Hence whoever is truly humble continually lies low in the knowledge of himself, and is afraid of nothing more than of being exalted.

Humility has been compared to a river which has a great deal of water in winter and scarce any at all in summer; for humility usually decreases in prosperity and increases in adversity.

Moral virtues are not to be acquired, any more

than arts and sciences, but by exercise and practice. To be a good artist, a good musician, a good orator, or a good philosopher, you must exercise yourself in the actions proper to each of these professions; so, to acquire humility, and other moral virtues, you must practice the arts belonging to them. It is true that all virtue must come from the hand of God; but it is true also that the same God, without whom we can do nothing, will have us also to cooperate with him. Humiliation is the way to humility, as patience is to peace of mind, and study to learning; if we will acquire humility we must put ourselves into the way of humility.

Many teachers of the spiritual life counsel us to take great heed lest we say anything which may turn to our own praise, which may make us pass for men of profound knowledge or eminent virtue. It is very hard for you to have any good quality that others will not perceive; if you take no notice of it yourself you will be the better loved for it and deserve a double praise; as well for being master of so good a quality as for being willing to conceal it. But if you make a show of it you will be laughed at. It is highly dangerous to take pleasure in hearing people praise and speak well of us. When we are praised we ought to cast our eyes upon our sins. Let us also be particular to take pleasure in hearing others praised. Whenever the good which you hear of your neighbor excites envy in you, or what you hear said of yourself causes self-satisfaction, be sure to look upon it as a fault. Do nothing to be seen and esteemed by men. Do not excuse yourself when in fault, for it is pride that makes us, as soon as we have committed one, or as soon as we are reproved for it, stand upon our defense. Prevent the imagination from indulging too freely in proud thoughts. Look on yourself as inferior to others, and prefer them to yourself.

Ought we to wish to be contemned? and, if we are, how shall we be able to bring forth fruit for the good of souls? For to make an impression by what we say, and to gain credit with an audience, we must be in esteem with them; so that on this account it seems even necessary to desire the esteem of men. The answer given by the fathers is that, though the great danger we incur by the honor and esteem of men ought to oblige us to avoid it, and though when we regard only ourselves we ought to wish to be despised, yet we may, nevertheless, with a view to God's greater glory, seek their approbation and esteem. It happens sometimes that good people rejoice at the good opinion which others have of them, but that is when they believe that thereby they can do more good to their souls; and then they do not so much rejoice at the esteem for themselves as at the benefit of their neighbor. For there is a great difference between seeking the applause of men and rejoicing at the salvation of souls. It is one thing to love the esteem of the world for its own sake, and to regard nothing but one's own satisfaction and the pleasure of glory, which is always wrong; and another thing to seek this esteem from a good motive, such as the advantage and salvation of your neighbor, which is very commendable. It is therefore permitted to desire the esteem of men, provided it be for the greater glory of God and their edification; for this is not to love one's own reputation. When one rejoices at the esteem of man it must be with such a regard only to God that at the very same moment that this esteem serves no further for God's glory and the salvation of our neighbor it ought rather to be a pain than a joy to us. For those who are thus disposed there is no fear when they accept any honor, or even speak to their own advantage, for they never do it but when they judge it necessary for the glory of God; and so the honor and praise which they receive leave no impression of vanity upon their heart.

That which makes the largest degree of humility so difficult to attain is that, on the one hand, we must use all imaginable care and diligence to acquire virtue, to resist temptations, and to be successful in all our pious undertakings, as if our own strength were sufficient to insure success; and, on the other hand, after having done all that depended on us, we are to confide no more in it than as if we had done nothing; we must look upon ourselves as unprofitable servants, and put our confidence in God alone.

There appears to be a conflict between humility and magnanimity; for magnanimity is a greatness of courage which urges us to undertake grand and glorious things, yet nothing seems more contrary to humility. The undertaking of great things seems wholly repugnant to humility, because this virtue demands that we acknowledge ourselves unworthy of everything and good for nothing; and it is presumptuous to attempt what we are not capable of performing. Also to attempt things which entitle us to honor seems acting still further against humility, because he who is truly humble ought to be far from so much as thinking how to attain honor. But the conflict is rather in seeming than in reality. For the attempting great things belongs properly to none but to him who is truly humble. To attempt great things in our own strength would indeed be presumption. But it is only upon diffidence in ourselves and confidence in God that Christian magnanimity lays the foundation of great enterprises; and humility does the same. There is nothing that we cannot do with the help of God. So with regard to honor. The magnanimous man desires only to deserve the glory without caring to possess it. He has raised himself so high above the opinion of the world that he finds nothing estimable but virtue; and, looking with the same eye upon the praise and scorn of men, he does nothing for the love of the one or for fear of the other.

As a man who has borrowed a great sum feels his joy for having the money alloyed by the obligation to restore it, and by the anxiety he is in as to whether he shall be able to pay it at the time appointed, so he who is humble, the more gifts he receives from God the more he acknowledges himself a debtor and under a stricter obligation to serve him; and, thinking that his gratitude and services do not answer, as they ought, the greatness of the favors and benefits he has received, he believes at the same time that anyone but himself would have made a better use of them. It is this which makes the servants of God more humble than others; for they know that God will call them to account not only for the sins they commit, but also for the benefits they receive. "Unto whomsoever much is given, of him shall be much required" (Luke xii, 48).

But why is God so pleased to exalt the humble, and to confer upon them so many favors? It is because all the good he does them returns to himself. For they who are humble appropriate to themselves nothing of what they receive; they restore it all to God, and, acknowledging that there is nothing great but the power of God alone, ascribe to him the glory and honor of all.

FRANCIS OF SALES.

It is very difficult to write briefly concerning so wonderful and lovable a man as Francis of Sales, both on account of his admirable traits and also on account of the abundant materials which have come down to us. There is an excellent life of him by Robert Ornsby, M.A., and a large volume of *memorabilia* entitled *The Spirit of St. Francis de Sales,* by his intimate friend, Bishop Camus, which is worthy to rank with Boswell's *Johnson.* Certain it is that few men have seemed so nearly perfect. It has been well said: "All things that command respect and attract love were found in Francis—high rank, polish of manners, geniality of disposition, shrewdness of head, vivacity of imagination, a capacity for profound theological studies, a rare felicity in the use of language, a captivating grace of manner, an almost unrivaled power as a director of souls, activity without bustle, mortification without sadness. There appears in his mind that union of sweetness and strength, of masculine power and feminine delicacy, of profound knowledge and practical dexterity, which constituted a character formed at once to win and subdue minds of almost every type and age."

He was born the oldest son of one of the principal nobles of Savoy, in the town of Sales, 1567. At the age of thirty-five he became bishop of Geneva, but his residence was at Annecy. After twenty years full of holy labors in this capacity he departed to glory, 1622.

In the year 1608 he issued the work by which he is best known, *The Introduction to a Devout Life.* It was drawn up chiefly from letters which he had written to one who was under his instruction, and which were so much admired in manuscript as to make their publication a necessity. The book immediately obtained a vast circulation throughout Europe, and its popularity has not waned down to the present day. Dr. E. M. Goulburn, himself one of the best spiritual writers of our own time, says: "There is no manual of devotion so winning, so attractive, and of such universal applicability as this. In profusion of imagery he is a very Jeremy Taylor. A man must be either the victim of inveterate sectarian prejudice or a stickler for the most vulgar theological commonplaces, or—much worse than either—dead to spiritual emotion, who can read Francis's treatise without a drawing of the heart toward its author, a longing after the devout life which he recommends, and a desire to act upon his instructions for leading it."

In 1616 he brought to completion what is in some respects his greatest work, the most profound, elaborate, and exhaustive, *A Treatise on the Love of God.* It is a mine of rich and beautiful thoughts, but not perhaps so generally useful to the ordinary reader as the *Introduction.*

Another volume from which we have made extracts in the following pages is entitled *Practical Piety,* and is composed of selections from Francis's letters and discourses. It is an admirable manual of devotion, treating very wisely of our duties toward God, toward our neighbor, and toward ourselves, as well as of the principal exercises of piety and the principal feasts of the year.

While he was emphatically the apostle of sweetness and gentleness, he had a dignity and gravity before which people stood in awe, and he had a burning hatred of sin as well as an ardent love for God. He did all things "passing well," but without vehemence, combining with intensity of devotion great calmness of spirit. He was hostile to anything like haste or flurry, overeagerness or anxiety. His favorite word was *pedetentim,* "by degrees," "step by step," "soon enough if well enough," not an inch in advance of God's will. He paid special attention to doing kindnesses for individuals, even the humblest, and if anyone treated him harshly he took particular pains to do him a favor. His passage from earth to heaven, though attended with intense pain, was most edifying. Exhortations to those around him to love God more were frequent. As some one gave expression to the thought of how necessary to the people his longer tarrying seemed to be, he replied, "A useless servant, useless, useless." The name of Jesus was the last word on his lips. His end was peace, and his works have certainly followed him. The few of his words here given are no better, perhaps, than much more which might be quoted, but they will afford a taste of his distinctive teaching.

THE MORNING EXERCISE.

1. Adore God most profoundly, and return him thanks for having preserved you from the dangers of the night; and if during the course of it you have committed any sin, implore his pardon. 2. Consider that the present day is given you in order that you may gain the future day of eternity; make a firm purpose, therefore, to employ it well with this intention. 3. Foresee in what business or conversation you will probably be engaged; what opportunities you will have to serve God; to what temptations of offending him you will be exposed, either by anger, by vanity, or any other irregularity; and prepare yourself by a firm resolution to make the best use of those means which shall be offered you to serve God and advance in devotion; as also, on the other hand, dispose yourself carefully to avoid, resist, and overcome whatever may present itself that is prejudicial to your salvation and the glory of God. 4. This done, humble yourself in the presence of God, acknowledging that of yourself you are in-

capable of executing your resolutions either to avoid evil or to do good; and, as if you held your heart in your hands, offer it, together with all your good designs, to his divine Majesty, beseeching him to take it under his protection, and so to strengthen it that it may proceed prosperously in his service.

THE EVENING EXERCISE.

Prostrate yourself before God, and recollect yourself in the presence of Jesus Christ crucified. Give thanks to God for having preserved you during the day past. Examine how you have behaved yourself throughout the whole course of it; and to do this more easily consider where you have been, with whom, and in what business you have been employed. If you find that you have done any good, thank God for it. If, on the other hand, you have done any evil, whether in thought, word, or deed, ask pardon of his divine Majesty, firmly resolving to confess it at the first opportunity, and to avoid it for the future. Recommend to the protection of divine Providence your soul and body, the holy Church, together with your parents and friends; and finally beg the Lord to watch over you. Thus, with the blessing of God, you may go to take that rest which his will has appointed for you.

TEMPTATION.

There are three steps to ascend to iniquity: temptation, delectation, and consent. Though the temptation to any sin whatsoever should last during life, it could never render us disagreeable to the divine Majesty provided that we were not pleased with it, and did not give our consent to it. The reason is, because we do not act, but suffer in temptation; and as in this we take no pleasure, so we cannot incur any guilt. It is not always in the power of the soul not to feel the temptation, though it be always in her power not to consent to it; it cannot hurt us so long as it is disagreeable to us. But with respect to the delectation which may follow the temptation it must be observed that, as there are two parts in the soul, the inferior and the superior, and the inferior does not always follow the superior, but acts for itself apart, it frequently happens that the inferior part takes delight in the temptation without the consent, nay, against the will, of the superior. This is that warfare which the apostle describes (Gal. v, 17) when he says that the flesh lusts against the Spirit, and that there is a law of the members and a law of the Spirit.

Therefore, whenever you are tempted to any sin, consider whether you have not voluntarily given occasion to the temptation; for then the temptation itself puts you in a state of sin, on account of the danger to which you have exposed yourself. When the delectation which follows temptation might have been avoided, and yet was not, there is always some kind of sin, more or less considerable, according to the time you have dwelt upon it or the pleasure you have taken in it.

As soon as you perceive yourself tempted follow the example of children when they see a wolf or a bear in the country; for they immediately run into the arms of their father or mother, or at least they call out to them for help. Look not the temptation in the face, but look only on our Lord; for if you look at the temptation, especially while it is strong, it may shake your courage. Divert your thoughts to some good and pious reflections, for when good thoughts occupy your heart they will drive away every temptation and suggestion.

It is a very good sign that the enemy keeps knocking and storming at the gate, for it shows that he has not what he wants. If he had he would not make any more noise, but enter in and quietly remain there.

FASTING.

We are greatly exposed to temptations, both when our body is too much pampered and when it is too much weakened, for the one makes it insolent with ease, and the other desperate with affliction.

Labor, as well as fasting, serves to mortify and subdue the flesh. Now, provided the labor you undertake contributes to the glory of God and your own welfare, I would prefer that you should suffer the pain of labor rather than that of fasting. Some find it painful to fast, others to serve the sick or visit prisoners, others to hear confession, to preach, to pray, and to perform similar exercises. These last pains are of more value than the former, for besides subduing the body they produce fruits much more desirable, and therefore, generally speaking, it is better to preserve our bodily strength more than may be necessary, in order to perform these functions, than to weaken it too much; for we may always abate it when we wish, but we cannot always repair it when we would.

In indifference respecting our food consists the perfection of the practice of that sacred rule, "Eat that which is set before you." I except, however, such meats as may prejudice the health or incommode the spirit, such as hot and high-seasoned meats; as also certain occasions in which nature requires recreation and assistance in order to be able to support some labor for the glory of God. A continual and moderate sobriety is preferable to violent abstinences practiced occasionally and mingled with great relaxations.

I think it a point of virtue to retire to rest early in the evening, that we may be enabled to awake and rise early in the morning, which is certainly, of all times, the most favorable, the most agreeable, and the least exposed to disturbance and distractions; when the very birds invite us to awake and praise God; so that early rising is equally serviceable to health and holiness.

CONVERSATION.

Let your language be meek, open, and sincere, without the least mixture of equivocations, artifice, or dissimulation; for although it may not be always advisable to say all that is true, yet it is never allowable to speak against the truth.

No artifice is so good and desirable as plain dealing. Worldly prudence and artifice belong to the children of the world; but the children of God walk uprightly, and their heart is without guile. Lying, double-dealing, and dissimulation are always signs of a weak and mean spirit.

In order to avoid contention do not contradict anyone in discourse, unless it be either sinful or very prejudicial to agree with him. But should it be necessary to contradict anyone, or oppose our own opinion to his, we must do it with much mildness and dexterity, so as not to irritate his temper; for nothing is ever gained by harshness and violence.

To speak little, a practice so much recommended by all wise men, does not consist in uttering few words, but in uttering none that are unprofitable; for in point of speaking one is not to regard the quantity so much as the quality of the words. But, in my opinion, we ought to avoid both extremes. For to be too reserved, and refuse to join in conversation, looks like disdain or a want of confidence; and, on the other hand, to be always talking, so as to afford neither leisure nor opportunity to others to speak when they wish, is a mark of shallowness and levity.

EVIL SPEAKING.

Rash judgment engenders uneasiness, contempt of our neighbor, pride, self-complacency, and many other most pernicious effects, among which detraction, the bane of conversation, holds the first place. Detraction is a kind of murder; for we have three lives, namely, the spiritual, which consists in the grace of God; the corporal, which depends on the soul; and the civil, which consists in our good name. Sin deprives us of the first, death takes away the second, and detraction robs us of the third. But the detractor by one blow of his tongue commits three murders: he kills not only his own soul and the soul of him that hears him, but also, by a spiritual murder, takes away the civil life of the person detracted. For, as St. Bernard says, both he that detracts and he that hearkens to the detractor have the devil about them, the one in his tongue and the other in his ear. As the serpent's tongue is forked and has two points, so is that of the detractor, who at one stroke stings and poisons the ear of the hearer and the reputation of him against whom he is speaking.

One act alone is not sufficient to constitute a vice. To acquire the name of a vice or a virtue the action must be habitual; one must have made some progress in it. It is then an injustice to say that such a man is passionate, or a thief, because we have seen him once in a passion or guilty of stealing. Also, since the goodness of God is so immense that one moment suffices to obtain and receive his grace, what assurance can we have that he who was yesterday a sinner is not a saint to-day? We can then never say a man is wicked without exposing ourselves to the danger of lying. All that we can say, if we must speak, is that he did such bad actions, or lived ill at such a time, that he does ill at present; but we must never draw consequences from yesterday to this day, nor from this day to yesterday, much less to to-morrow.

Some, to avoid the sin of detraction, commend and speak well of vice. We must avoid this extreme. We must openly blame that which is blamable; for in doing this we glorify God, provided we observe the following conditions. To speak commendably against the vices of another it is necessary that we should have in view the profit either of the person spoken of or of those to whom we speak. It is, moreover, requisite that it should be my duty to speak on this occasion, as when I am one of the chief of the company; for if I should keep silence I would seem to approve of the vice; but if I be one of the least I must not take upon me to pass my censure. But above all it is necessary that I should be so cautious in my remarks as not to say a single word too much. My tongue, whilst I am speaking of my neighbor, shall be in my mouth like a knife in the hand of a surgeon, who would cut between the sinews and the tendons. The blow I shall give shall be neither more nor less than the truth. In fine, it must be our principal care in blaming any vice to spare as much as possible the person in whom it is found.

When you hear anyone spoken ill of make the accusation doubtful if you can do it justly; if you cannot, excuse the intention of the party accused; if that cannot be done, express a compassion for him and change the topic of conversation, remembering yourself, and putting the company in mind, that they who do not fall owe their happiness to God alone; recall the detractor to himself with meekness, and declare some good action of the party offended, if you know any.

QUIETNESS OF SPIRIT.

We ought above all things to secure our tranquillity, not only because it is the mother of contentment, but chiefly because it is the daughter of the will of God and of the resignation of our own will.

We shall soon be in eternity, and then we shall see what a little matter are all the affairs of the world, and of how small consequence it was whether they were done or not done. Nevertheless we now make ourselves anxious, as though they were great things. When we were little children with what earnestness did we gather bits of tiles, wood, and clay, to build little houses with, and when anyone destroyed them we were greatly distressed at it, and wept; but now we know right well that all that was of little consequence. We shall do the same in heaven one day, when we shall see that our interests in the world were all mere childishness. Let us pursue our childish occupations, since we are children, but let us not catch cold about them; and if anyone throws down our little houses and designs let us not be overdistressed; for when night comes—I mean death—and we must return to our homes, our little houses will all be useless. We must return to our Father's house.

It is a truth that nothing can give us a deeper tranquillity in this world than frequently to look upon our Lord in all the afflictions which came upon him from his birth until his death. For we shall there see so much scorn, calumny, poverty, need, abjection, pains, torments, injuries, and all sorts of bitterness, that, in comparison with it we find out that we were wrong in calling by the name of affliction, pain, and contradiction those little accidents which happen to us, and in desiring patience for such a trifling matter, since one little drop of modesty should amply suffice to support that which happens to us.

HUMBLE-MINDEDNESS.

"Borrow empty vessels not a few," said Elisha to the poor widow, "and pour oil into them." To receive the grace of God into our hearts they must be emptied of vainglory. We call that glory vain which we assume to ourselves either for what is not in us or for what is in us, and belongs to us, but deserves not that we should glory in it.

Generous minds do not amuse themselves about the petty toys of rank, honor, and salutation; they have other things to perform; such baubles only belong to degenerate spirits. He that may have pearls never loads himself with shells; and such as aspire to virtue trouble not themselves about honors. Everyone, indeed, may take and keep his own place without prejudice to humility, so that it be done carelessly and without contention. For as they that come from Peru, besides gold and silver, bring also thence apes and parrots, because they neither cost much nor are burdensome, so they that aspire to virtue refuse not the rank and honor due to them, provided it cost them not too much care and attention, nor involve them in trouble, anxiety, disputes, or contentions. Nevertheless I do not here allude to those whose dignity concerns the public, nor to certain particular occasions of important consequences; for in these everyone ought to keep what belongs to him with prudence and discretion, accompanied by charity and suavity of manners.

I would neither pretend to be a fool nor a wise man; for if humility forbids me to conceal my wisdom, candor and sincerity also forbid me to counterfeit the fool; and as vanity is opposite to humility, so artifice, affectation, and dissimulation are contrary to sincerity.

The best abjections, those most profitable to our souls and most acceptable to God, are such as befall us by accident or by our condition of life; because we have not chosen them ourselves but received them as sent by God, whose choice is always better than our own.

Humility not enduring that we should have any opinion of our own excellence, or think ourselves worthy to be preferred before others, cannot permit that we should seek after praise, honor, and glory, which are only due to excellence; yet she consents to the counsel of the wise man who admonishes us to be careful of our good name, because a good name is our esteem, not of an excellence, but only of an ordinary honesty and integrity of life, which humility does not forbid us either to acknowledge in ourselves or to desire the reputation of it. It is one of the foundations of human society, without which we are not only unprofitable but prejudicial to the public by reason of the scandal it would receive. Charity requires and humility consents that we should desire it and carefully preserve it.

The obligation of preserving our reputation, and

of being actually such as we are thought to be, urges a generous spirit forward with a strong and agreeable impulse. An excessive fear of losing our good name betrays a great distrust of its foundation, which is the truth of a good life. He that is too anxious to preserve his reputation loses it; and that person deserves to lose honor who seeks to receive it from those whose vices render them truly infamous and dishonorable.

That humility which does not produce generosity is undoubtedly false. For after humility has said, I can do nothing, I am nothing, it immediately gives place to generosity, which says, There is nothing which I cannot do, inasmuch as I put all my confidence in God, who can do everything. And with this confidence humility, consequently, undertakes everything which it is ordered to do, how difficult soever. And if it applies itself to fulfill the commandment in simplicity of heart, God will rather work a miracle than fail of giving it his aid; because it is not from any confidence in its own strength that humility undertakes the work, but from the confidence which it has in God.

Behold the example which we ought to follow when we are ordered to do anything. We ought to undertake it generously, without reckoning on ourselves, but reckoning much on the grace of God, who wills that we should obey without making any resistance. But I well understand the subtlety of false humility; it is that we fear we shall not come forth with honor to ourselves. We value our reputation so highly that in the exercise of our office we do not like to be reckoned as apprentices, but as masters who never commit any blunders at all.

It is a good practice of humility never to look upon the actions of our neighbors except to remark the virtues that are in them, never their imperfections; for so long as we are not in charge of them we must never turn our eyes, and still less our attention, on that side.

We must always put the best construction that we can upon what we see our neighbor do. In doubtful matters we ought to persuade ourselves that what we noticed is not bad, but that it is our imperfections that cause such a thought to arise in our minds; that thus we may avoid rash judgments, a very dangerous evil, for which we ought to have a sovereign detestation.

To acquire the spirit of humility there is no other way but frequent repetition of its acts. Humility makes us annihilate ourselves in all those things which are not necessary for our advancement in grace, such as good speaking, noble mien, great talents for the management of affairs, a great spirit of eloquence, and the like; for in these exterior things we ought to desire that others should succeed better than ourselves.

Love your abjection. That is, remain humble,

tranquil, sweet, full of confidence in the midst of this obscurity; do not make yourself impatient, or trouble yourself for all this, but with a good heart —I do not say gayly, but I do say freely and firmly—embrace this cross, and remain under these clouds. Love to be obscure, for the love of Him who wishes you to be so, and you will love your own abjection.

HOLY INDIFFERENCE.

It is difficult to give an exact definition of the holy indifference of a will dead to itself and totally absorbed in the will of God. According to my idea of a perfectly indifferent soul, which desires nothing, and permits the Almighty to will whatever he pleases, it should be defined as a will in a state of simple and general expectation, disposed for all events. Yet, though the expectation of the soul is a simple disposition to receive whatever may occur, not an action, it is still perfectly voluntary. After these events have happened expectation is changed into consent or acquiescence. Before they occur it is simple expectation; that is, a disposition of the soul by which she is prepared for everything, and perfectly indifferent as to whatever it may please the divine will to ordain.

To exercise persons in this holy indifference God sometimes inspires them with very exalted designs, which are not meant to succeed. Their duty on these occasions is, on the one side, to commence with a noble courage and simple confidence, and to persevere with constancy, as long as a hope of success remains; and, on the other hand, tranquilly and humbly to accept whatever degree of success God is pleased to give their exertions. Happy are the souls in whom God discovers this perfect readiness to abandon, by his desire, the enterprises which they have generously and courageously undertaken in obedience to his commands. Nothing more clearly proves perfect indifference than to abandon the execution of a good design when God pleases that it should succeed no further. It was God who urged us onward and served as a guide; we advanced with ardor for his glory, and at the first intimation of his will we unhesitatingly retraced our steps.

The absolute will of God is usually known only by the event which is its effect. Before this takes place we should unite ourselves to the divine will which is called signified; and when this adorable will is made manifest in after occurrences we should immediately attach ourselves thereto by amorous submission. Let us suppose that I, or some one very dear to me, have been attacked by serious illness; does God will that the malady should or should not be followed by death? This I do not, and cannot, discover. But I know by his signified will that he requires me, while waiting for the event which he has ordained, to employ the remedies necessary for my recovery. I shall then use them, and omit nothing calculated to remove my illness. But if it be the will of God that the remedies prove ineffectual, and the sickness terminate in death, as soon as I shall have discovered this to be the will of God the superior part of my soul will cheerfully submit, notwithstanding the repugnances of the inferior part.

Everything which occurs in the universe, except sin, happens by the will of God, which is called absolute or of good pleasure; no one can prevent its accomplishment, and it is known by the effects it produces. When events occur we judge unhesitatingly that God has willed and regulated them.

But, you will object, when an enterprise inspired by God fails, through the fault of the person to whom it has been committed, how can he then acquiesce in the divine will, knowing that it is not the will of God which has prevented success, since it is not, and cannot be, the cause of the fault which has impeded the happy termination of the enterprise? Your fault certainly does not proceed from the will of God, because God cannot be the author of sin; yet it is his will that your fault be punished by the failure of the undertaking. As he is infinitely good he cannot will sin which offends him; but as his justice is no less infinite than his goodness he wills the punishment which is the consequence of your fault.

Thus should we act: our will should be as easily molded by the will of God as soft wax is shaped by the hand; we should not amuse ourselves in forming desires and projects; we should have no views or pretensions, but leave the dispòsal of everything belonging to us in the hands of God. Let us bless and thank God on all occasions, saying, I do not wish for anything, O my God; I do not even desire to know what may befall me; the power of willing and choosing belongs to thee; I reserve to myself only that of blessing thee for whatever thou hast ordained. How excellent a use do we make of our liberty when, suppressing all desires and natural solicitude, we are solely occupied in praising the divine will, which regulates all things, and blessing its ever-equitable decrees.

The brilliancy of the stars is not obscured when the sun enlightens our horizon, but it is concealed from us by the light of the sun, and seems to be engulfed in that immense ocean of splendor in which it is lost. In like manner the human will is not destroyed when it abandons itself totally to that of God; but it is so absorbed in the divine will that it cannot be distinguished from it, having in reality no effect, no desire, no will, but the will of God.

When a servant who follows his master is asked where he goes he might reply that he does not go, he only follows; because it is his master's will, and not his, which determines the place to which he walks. The will when totally abandoned to that of God desires nothing according to its own choice; it simply follows the selection made by the Almighty.

To sail is not to proceed by our own motion, but by that of the vessel in which we have embarked. The human mind may be said to embark when it abandons itself to the will of God, allowing itself to be conducted by this adorable will, to receive its motion and not to move itself. It is like an infant at the breast, which, being unable to dispose of itself, has no will except to love its mother; on whichever side it is placed it is satisfied, provided it be in the arms of her whom it loves and with whom it seems to constitute but one object. As it is not aware of having a will it does not make any exertion to unite it to its mother's, but it abandons itself to her care and allows her to will whatever she pleases in its regard. Souls thus united to God have reached the highest degree of perfection which can be attained in this life.

WHEN IS LOVE TO GOD MOST PERFECT?

A heart inflamed with divine love adores and loves the will of God, not only in the consolations it imparts, but also in the afflictions it is pleased to send; it even loves it more ardently in crosses and trials than in consolations, because the peculiar effect of a strong and generous love is to suffer for the object of predilection.

To love the will of God in the consolations it sends us is a real and sincere love, provided it be indeed the will of God we love in his consolations, and not the consolations in his divine will. This is, however, a species of love which has no efforts to make, no contradictions or repugnances to surmount. For who would not love so amiable a will under circumstances so gratifying to nature? To love the will of God in its commandments, in its counsels and inspirations, is a second degree of love, much more perfect than the first; because it leads us to renounce our own will and to deprive ourselves of many pleasures, though it does not forbid them. To love the will of God in sufferings and afflictions is the third and sovereign degree of charity. Under these circumstances we can discover nothing amiable but the divine will itself; we experience great natural repugnance, and not only renounce pleasure, but even embrace sufferings and pains. Divine love is always fearful when it seeks the will of God amid consolation; because it is easy, under these circumstances, to love our own happiness rather than the divine will. But we practice the highest perfection of love when we not only receive afflictions with patience and resignation, but even cherish and delight in them on account of the will of God from which they proceed.

If I only wish for clear water it is of little consequence whether it be brought in a vase of gold or of glass. I should even receive it with more pleasure when presented in a glass, because I can then see it more clearly than in a golden cup. In like manner, if I seek only the will of God I should be indifferent whether it be presented to me in tribulation or consolation, provided I can clearly discern it. It should even be more agreeable in suffering, because it is then more visible; and the only amiability of tribulation is that which it borrows from the divine will.

To comprise all in a few words, the divine will is the sovereign object and ruling attraction of a soul influenced by holy indifference. Wherever she can discover the divine will she eagerly unites herself thereto; and amidst several objects, all marked with the seal of God's adorable will, that in which this will is most evidently manifested always receives her preference, whatever motives may incline her to the contrary. The divine will sweetly conducts the indifferent soul as it pleases.

The love of our relatives, benefactors, and friends is in itself very conformable to the will of God; but it ceases to be so when it becomes excessive. Souls which are inordinately attached to the objects which God wills they should love may certainly be said to love God above all things; yet they do not love him in all things, since their affection for many objects is founded on other motives in which God has no share, though they are not opposed to his divine will.

That soul is the most cherished by her heavenly spouse, and the most ardent among his sacred lovers, who not only loves God above all things and in all things, but who loves God alone in all things; who, to speak more correctly, amidst many things which are the objects of her affection loves but one, which is God. A certain proof that we love God alone in all things is when we love him equally under all circumstances; because, God being always the same immutable Being, the inequality of our love for him can only proceed from a particular motive produced by the consideration of some object which is not God.

LITTLE THINGS.

We have not always an opportunity of doing great things; but we can hourly perform insignificant actions with an ardent love. To conform to the different characters of the persons with whom we associate; to bear their disagreeable and unpolished manners, which annoy and revolt, consequently to gain frequent victories over our passions and inclinations; to contradict our natural aversions; to conquer our antipathies; to acknowledge our faults, and to receive with humility the confusion resulting from them; to correct the natural variabilities of temper and be continually on our guard against the obstacles which oppose the peace of our souls; to love abjection, and joyfully to receive the contempt and censure incurred by our manner of life, conduct, and actions—all this, when embraced through love and animated by holy dilection, contributes more than we are aware to our spiritual advancement.

The most trivial actions are performed with great merit when accompanied with purity of intention and an ardent desire to please God. Some devout persons perform many good works without advancing much in charity, because they do everything tepidly, and act more from natural inclination than by the inspiration and emotion of grace. Others, on the contrary, to judge by the exterior of their actions, do very little for God; but this little is accompanied with so much purity of intention that their progress in holy dilection is rapid.

SEEK PERFECTION SENSIBLY.

You perhaps think that perfection is to be found ready-made, and that you only require to put it on, as you would put on a garment; but it is not so; it is necessary to make it yourself, and to clothe yourself with it.

You seem to think that perfection is an art, and that if one could find out the secret of it one would have it without any trouble. Certainly we deceive ourselves; for there is no other nor greater secret than to do and to labor faithfully in the exercise of divine love if we wish to unite ourselves unto the beloved.

Take care to make yourself daily more pure in

heart; this purity consists in weighing everything in the balance of the sanctuary, which is nothing else than the will of God.

Know that it is the virtue of patience that insures us the most perfection; and if we must have it with others we must also have it with ourselves. Those who aspire to the pure love of God have not so much need of patience with others as with themselves.

Keep your eyes lifted up unto God. Augment your courage in holy humility; fortify it in sweetness; confirm it in evenness. Make your spirit perpetually the master of your inclinations and humors. Never allow apprehensions to enter into your heart. Each day will give you the knowledge of what you shall best do the next.

As much as you can, do perfectly that which you do; but when it is done do not think any more about it; think of what is to be done next. Walk very simply with the cross of our Lord, and do not torment your mind. We ought to hate our defects; but with a tranquil and peaceful hatred, not with a troubled and distempered hatred. And, further, we ought to have patience when we see them, and derive from them the profit of a holy abasement of ourselves. We must be charitable with our soul and not devour it when we see that it does not err with its full consent. Do not lose courage, have patience, wait, exercise yourself strongly in the spirit of compassion; I do not doubt but that God will hold you with his hand.

It appears to me that our faults universally proceed from no other cause but this, namely, that we forget the maxim of the saints who have warned us that we ought every day to consider that we are commencing anew our advancement in our perfection. The work is never finished; it must always be recommenced, and recommenced with a good heart. What we have done up to the present time is good, but what we are about to begin shall be better; and when we shall have finished we will recommence something else which shall be still better; and then again something else, until we go out of this world to commence another life, which shall have no end because nothing better can happen to us.

Do not then examine so carefully whether you are in perfection or not; here are two reasons why you should not. One is that it is to no purpose our examining ourselves in this way, since, were we the most perfect souls in the world, we ought never to know or be aware of it, but to esteem ourselves always as imperfect. The other reason is that this examen, when it is made with anxiety and perplexity, is only a loss of time; and those who make it are like musicians who make themselves hoarse with practicing a motet; for the mind wearies itself with an examen so great and so continual, and when the time of execution arrives it can do no more.

Simplify your judgment; do not make so many reflections and replies, but go on simply and with confidence; for you there is nothing else in the world but God and yourself. You have nothing to do with aught else, except so far as God commands it and in the way in which he commands it to you. Avoid minutely examining what other people do, or what will become of them; look on them with an eye simple, good, sweet, and affectionate. Do not require in them more perfection than in yourself, and do not be astonished at the diversity of imperfections; for imperfection is not greater imperfection merely because it is unusual. Behave like the bees—suck the honey from all flowers and all herbs. Go on joyously, and with open heart, as much as you can; and if you do not always go on joyously, at least go on always courageously and confidently.

VIRTUE TESTED.

When people say to me, Look at such a sister, in whom one sees no imperfection, I immediately ask, Does she hold any office? If they say not, then I make no great account of her perfection. For there is a great difference between the virtue of this sister and that of another who shall be well tried, whether interiorly by temptations, or exteriorly by contradictions; for the virtue of strength and the strength of virtue are not ordinarily acquired so perfectly in time of peace as they are whilst we are tried by temptation of its contrary. There is a great deal of difference between the absence of a vice and the presence of the opposite virtue. Many appear to be highly endowed with virtue who, nevertheless, are not so, because they have not acquired it by labor. We ought always to remain humble, and not to suppose that we have the virtues merely because we do not commit, or at least do not know that we commit, the faults opposed to them.

It is a maxim of marvelous efficacy, "Let God put me in what service he wills, 'tis all one to me, provided that I serve him!" But take care to chew it well over and over in your mind; make it melt in your mouth, and do not swallow it whole. St. Teresa says somewhere that we very often say such words from habit, and a certain slight idea of them, and we fancy that they are spoken from the deep of our heart, although it was nothing of the sort, as we afterward discover by our practice.

THE FEVER OF SELF-WILL.

We like to serve God according to our own will, and not according to his. God commands me to save souls, and I wish to remain in contemplation; the contemplative life is good, but not to the prejudice of obedience. It is not for us to choose according to our will; we must will what God wills; and if God wills that I should serve him in one capacity I must not will to serve him in another.

There is no vocation which has not its annoyances, bitternesses, and vexations; and much more, if we except those who are fully resigned to the will of God, each person would willingly change his condition for that of others. Whence comes this general disquietude of minds unless from a certain dislike which we all have to constraint? But it is all one. Whoever is not fully resigned, he may turn to this side or to that, he will never find repose. Those who have a fever find no place to their mind. They have not remained a quarter of an hour in one place when they would be in another. It is not the bed that causes their restlessness, but the fever which torments them everywhere. A person who has not the fever of self-will is contented everywhere provided that God is served. Such a one does not trouble himself about what capacity God employs him in; provided that he does his divine will it is to him all one.

Often reflect that all we do derives its true value from the conformity which we have to the will of God; so that in eating and drinking, if I do it because it is the will of God that I do it, I am more pleasing to God than if I suffered death without that intention.

LOVE OF OUR OWN OPINION.

Everyone has opinions of his own. What we must avoid is attaching ourselves to them and loving them; because that attachment and that love are very contrary to perfection. The love of our own judgment, and the value which we set on it, is the cause why there are so few perfect souls. When we are required, either by charity or obedience, to give our advice on a subject that is under discussion, we must do it simply, making ourselves, for the rest, indifferent whether it is received or not. The matter being decided, we must say no more about it, especially with those who were of our way of thinking; for that would be to nourish this defect, and to show that we have not completely submitted to the advice of the others, and that we always prefer our own. We must not even think about it any more, unless the resolution taken is remarkably faulty; for in that case if any means could still be found to prevent its execution, or to apply a remedy to it, we ought to adopt such means in the most charitable and quiet way we can, so as not to trouble anyone, or to bring into contempt what they thought good. The love of our own opinion is the last thing that we part with; and nevertheless it is one of the most necessary to part with for the acquisition of true perfection; for otherwise we do not acquire holy humility, which forbids us from making any account of ourselves or of anything that depends upon us.

There are souls who will not, as they say, be led except by the Spirit of God. And they fancy that all the things they imagine are so many inspirations and movements of the Holy Ghost, who takes them by the hand and conducts them like children in all that they would do. In this they greatly deceive themselves. For is there any vocation more marvelous than that of St. Paul, in which our Lord himself spoke to him in order to convert him? and nevertheless he would not instruct him, but sent him to Ananias to learn whatever he had to do. And although St. Paul might have said, "Lord, wherefore not thyself?" he did not say so, but went in all simplicity to do what was commanded him. After this, shall we think ourselves more favored of God than St. Paul, and believe that he wills to conduct us himself without the instrumentality of any creature?

SWEETNESS OF TEMPER.

Let us be very sweet and humble in heart toward all, but above all toward our own. Let us not agitate ourselves; let us go on with all sweetness, bearing with one another. Let us take good care that our heart does not escape us.

Generous devotion does not wish to have companions in everything it does, but only in its aim, which is the glory of God and the advancement of our neighbor in divine love. And provided that it goes straight to that end it does not trouble itself by what road. Provided that he who fasts fasts for God, and that he who fasts not also for God fasts not, it is as content with the one as with the other. Generous devotion does not wish to attract others to its own mode of life, but it follows its own path simply, humbly, tranquilly. If this point be well understood and well observed it will preserve in souls a marvelous tranquillity of mind and a great sweetness of heart. Let Martha be active, but let her not control Mary. Let Mary be contemplative, but let her not despise Martha. For our Lord will take up the defense of her who is censured.

Blessed are the pliable hearts, for they will never break. The effects of true liberty are a great sweetness of spirit, a great gentleness and readiness to yield wherever there is not sin or danger of sin. It creates a disposition sweetly pliable in the action of all virtue and charity. For example, a soul that has attached itself to the exercise of meditation; interrupt it, and you will see it lay aside that exercise with some expression of annoyance, disturbed and put out. A soul which has true liberty will lay aside its meditation with an even countenance and a heart graciously disposed toward the troublesome person who may have caused it inconvenience. For to such a soul it is all one whether it serves God by meditating or serves him by bearing with its neighbor. Both the one and the other are the will of God; but to bear with its neighbor is necessary at that particular moment. The occasions of this liberty are all things which occur contrary to our inclination; for whoever has not his inclinations fixed is not disquieted when they meet with opposition.

PATIENCE.

Limit not your patience to this or that kind of injuries and afflictions, but extend it universally to all those that it shall please God to send you. He that is truly patient suffers indifferently tribulations whether accompanied by ignominy or honor. To be despised, reprehended, or accused by wicked men is pleasant to a man of good heart; but to suffer blame and ill treatment from the virtuous, or from our friends and relations, is the test of true patience. The evils we suffer from good men are much more insupportable than those we suffer from others; and yet it often happens that two good men, having each of them the best intentions, through a diversity of opinion foment great persecutions and contradictions against each other.

We must not only bear sickness with patience, but also be content to suffer sickness under any disorder and in any place, amongst those persons and with those inconveniences which God pleases; and the same must be said of other tribulations. When any evil befalls you apply the remedies that may be in your power agreeably to the will of God; for to act otherwise would be to tempt divine Providence. Having done this, wait with resignation for the success it may please God to send; and should the remedies overcome the evil, return him thanks with humility; but if, on the contrary, the evils overcome the remedies, bless him with patience.

Whenever you are justly accused of a fault humble yourself, and candidly confess that you deserve more than the accusation which is brought against you; but if the charge be false excuse yourself meekly, denying your guilt; for you owe this respect to truth and to the edification of your neighbor. But if, after your true and lawful excuse, they should continue to accuse you, trouble not yourself, nor strive to have your excuse admitted; for, having discharged your duty to truth, you must also do the same to humility, by which means you neither offend against the care you ought to have of your reputation nor the love you owe to peace, meekness of heart, and humility.

Complain as little as possible of the wrongs you suffer; for, commonly speaking, he that complains sins, because self-love magnifies the injuries we suffer and makes us believe them greater than they really are. The truly patient man neither complains himself nor desires to be pitied by others. He speaks of his sufferings with truth and sincerity, without murmuring, complaining, or aggravating the matter. He patiently receives condolence, unless he is pitied for an evil which he does not suffer; for then he modestly declares that he does not suffer on that account; and thus he continues peaceable betwixt truth and patience, acknowledging but not complaining of the evil.

GOOD THOUGHTS FROM EVERYTHING.

One may extract good thoughts and holy aspirations from everything that presents itself amidst the variety of this mortal life. A devout soul standing over a brook on a very clear night, and seeing the heavens and stars therein represented, exclaimed, "O my God, these very stars which I now behold shall be one day beneath my feet, when thou shalt have lodged me in thy celestial tabernacles; and as the stars of heaven are here represented, even so are the men of this earth represented in the living fountain of divine charity." Another, seeing a river flowing swiftly along, cried out, "My soul shall never be at rest till she be swallowed up in the sea of the divinity, her original source." Another, contemplating a pleasant brook, upon the bank of which she was kneeling at her prayers, being rapt into an ecstasy, often repeated these words, "The grace of my God flows thus gently and sweetly, like this little stream." Another, looking on the trees in bloom, sighed and said, "Ah, why am I alone without blossoms in the garden of the Church!" Another, seeing little chickens gathered together under the hen, said, "Preserve us, O Lord, continually under the shadow of thy wings." Another, looking upon the flower called heliotropium, which turns to the sun, said, "When shall the time come, O my God, that my soul shall faithfully follow the attractions of thy goodness?" And seeing the flowers called pansies, which are beautiful but without fragrance, "Ah!" said he, "such are my conceptions; fair in appearance, but of no effect, producing nothing."

As the great work of devotion consists in the exercise of spiritual recollection and ejaculatory prayers, the want of all other prayers may be supplied by them; but the loss of these can scarcely be repaired by any other means. Without them we cannot lead a good, active life, much less a contemplative one. Without them repose would be but idleness, and labor vexation. Wherefore I conjure you to embrace this; exercise your whole heart, without ever desisting from its practice.

"HOLY LIVING AND DYING."

THE RIGHT REV. JEREMY TAYLOR, D.D., Bishop of Down, Connor, and Dromore, who wrote the book named above, took an active part in stirring times. Born at Cambridge, England, in 1613, he entered college there in 1626, received holy orders in 1633, was made bishop in 1661, and died in 1667. Being an ardent royalist, he espoused the cause of King Charles in his struggle with the Parliament; hence, during the supremacy of the latter and the protectorate of Cromwell, he suffered considerable persecution, and was several times imprisoned. At the restoration of the monarchy in 1660 he properly came in for his share of the favors distributed.

He was an eloquent preacher and a very saintly man, presenting, it has been said, "as fine a pattern of a Christian bishop as the annals of the Church of England afford." Nature did much for him, and grace still more. His manners were gentle, his humility was deep, his charity boundless, while he had an acute and vigorous mind, as well as extensive learning and much practical wisdom. He has been called "the Homer of divines," "the Shakespeare of the Church," and "the Spenser of English theological literature." It is through his writings that he chiefly lives. Many of his printed sermons show great powers of thought, as well as an exuberant imagination. He composed numerous books, prominent among which are *The Liberty of Prophesying,* in which he nobly advocates the widest principles of toleration; *The Great Exemplar of Sanctity,* a popular life of Christ; and *Ductor Dubitantium* (or *The Rule of Conscience*), a large work in two volumes, on which he founded his brightest hopes of renown and usefulness. These hopes, however, were disappointed. He is mainly known to-day by his *Holy Living and Dying,* published originally (in 1650 and 1651) as two separate productions, *The Rule and Exercise of Holy Living,* and *The Rule and Exercise of Holy Dying.* It passed through nineteen editions within a little more than fifty years after publication, and is by far the most noted manual of devotion produced in the Church of England.

The learned and pious author, in his dedication or preface, says that he has been led "to draw into one body those advices which the several necessities of many men must use at some time or other, and many of them daily; that by a collection of holy precepts they might less feel the want of personal and attending guides, and that the rules for conduct of souls might be committed to a book which they might always have; since they could not always have a prophet at their needs, nor be suffered to go up to the house of the Lord to inquire of the appointed oracles." Such a design was most worthy, and it was most admirably carried out, with excellent results. John Wesley was greatly indebted to this book, as well as to à Kempis, perusing both at about the same time. After reading Taylor on purity of intention he says, "Instantly I resolved to dedicate all my life to God, all my thoughts, words, and affections, being thoroughly convinced there was *no medium*, but that every part of my life, not some only, must either be a sacrifice to God or myself, that is, in effect, to the devil."

A considerable part of the thick volume (515 12mo pages) is occupied with prayers. A long section in the "Holy Dying" comprises counsels to the "clergy-guides" for ministering to the sick, another section analyzes the decalogue "for the assistance of sick men in making their confessions to God and his ministers." In short, as in all such ancient books, it is only a small portion that can be set down as of perpetual value, adapted to all ages and lands. We do not, of course, in this chapter give the whole even of this small portion, but we do furnish, we think, the choicest of the thoughts. If they shall be seized upon by the reader with as vigorous an apprehension and as practical a purpose as John Wesley exercised, they will be sufficient to transform his life. There are very few guides to holy living better than good Jeremy Taylor.

PURITY OF INTENTION.

This grace is so excellent that it sanctifies the most common actions of our life; and yet so necessary that, without it, the very best actions of our devotion are imperfect and vicious. For, as to know the end distinguishes a man from a beast, so to choose a good end distinguishes him from an evil man. The praise is not in the thing done, but in the manner of its doing. If a man visits his sick friend, and watches at his pillow for charity's sake and because of his old affection, we approve it; but if he does it in hope of a legacy he is a vulture, and only watches for the carcass. The same things are honest and dishonest; the manner of doing them, the end of the design, makes the separation.

Holy intention is to the actions of a man that which the soul is to the body, or form to its matter, or the root to the tree, or the sun to the world, or the fountain to a river, or the base to a pillar; for without these the body is a dead trunk, the matter is sluggish, the tree is a block, the world is darkness, the river is quickly dry, the pillar rushes into flatness and a ruin; and the action is sinful, or unprofitable and vain.

In every action reflect upon the end; and in your undertaking it consider why you do it, and what you propound to yourself for a reward.

Let every action of concernment be begun with prayer, that God would not only bless the action but sanctify your purpose; and make an oblation of the action to God.

It is likely our hearts are pure and our intentions spotless when we are not solicitous of the opinion and censures of men, but only that what we do be our duty and accepted of God. For our eyes will certainly be fixed there from whence we expect our reward; and if we desire that God should approve us it is a sign we do his work, and expect him our paymaster.

He that does as well in private, between God and his own soul, as in public, in pulpits, in theaters, and market places, hath given himself a good testimony that his purposes are full of honesty, nobleness, and integrity. But he that would have his virtue published studies not virtue but glory. He is not just that will not be just without praise; but he is a righteous man that does justice when to do so is made infamous; and he is a wise man who is delighted with an ill name that is well gotten. And indeed that man hath a strange covetousness, or folly, that is not contented with this reward, that he hath pleased God.

It is well, also, when we are not solicitous or troubled concerning the effect and event of all our actions; but that being first by prayer recommended to Him is left at his dispose. For then, in case the event be not answerable to our desires, or to the efficacy of the instrument, we have nothing left to rest in but the honesty of our purposes; which it is the more likely we have secured by how much more we are indifferent concerning the success. If thou beest much troubled that thy labors prove unprosperous, it is certain thou didst not think thyself secure of a reward for thine intention; which thou mightest have done if it had been pure and just.

He loves virtue for God's sake and its own that loves and honors it wherever it is to be seen. But he that is envious or angry at a virtue that is not his own, at the perfection or excellency of his neighbor, is not covetous of the virtue, but of its reward and reputation; and then his intentions are polluted. It was a great ingenuity in Moses that wished all the people might be prophets; but if he had designed his own honor he would have prophesied alone. But he that desires only that the work of God and religion shall go on is pleased with it, whosoever is the instrument.

He that despises the world, and all its appendant vanities, is the best judge, and the most secured of his intentions, because he is the furthest removed from a temptation. Every degree of mortification is a testimony of the purity of our purposes; and in what degree we despise sensual pleasure, or secular honors, or worldly reputation, in the same degree we shall conclude our heart right to religion and spiritual designs.

When we are not solicitous concerning the instruments and means of our actions, but use those means which God hath laid before us with resignation, indifferency, and thankfulness, it is a good sign that we are rather intent upon the end of God's glory than our own conveniency or temporal satisfaction.

CARE OF OUR TIME.

He that is choice of his time will also be choice of his company and choice of his actions, lest the first engage him in vanity and loss, and the latter, by being criminal, be a throwing of his time and himself away, and a going back in the accounts of eternity.

God hath given every man work enough to do, that there shall be no room for idleness; and yet hath so ordered the world that there shall be space for devotion. He that hath the fewest businesses of the world is called upon to spend more time in the dressing of his soul, and he that hath the most affairs may so order them that they shall be a service of God; whilst at certain periods they are blest with prayers and actions of religion, and all day long are hallowed by a holy intention. So that no man can complain that his calling takes him off from religion; his calling itself, and his very worldly employment in honest trade and offices, is a serving of God; and if it be moderately pursued, and according to the rules of Christian prudence, will leave void spaces enough for prayers and retirements of a more spiritual religion.

In the morning, when you awake, accustom yourself to think first upon God, or something in order to his service; and at night also let him close thine eyes. And let your sleep be necessary and healthful, not idle and expensive of time beyond the needs and conveniences of nature; and sometimes be curious to see the preparation which the sun makes when he is coming forth from his chambers of the east.

Never talk with any man, or undertake any trifling employment, merely to pass the time away; for every day well spent may become a "day of salvation," and time rightly employed is an "acceptable time."

In the midst of the works of thy calling often retire to God in short prayers and ejaculations; and those may make up the want of those larger portions of time which, it may be, thou desirest for devotion, and in which thou thinkest other persons have advantage of thee.

Let not your recreations be lavish spenders of your time; but choose such which are healthful, short, transient, recreative, and apt to refresh you. But at no hand dwell upon them, or make them your great employment; for he that spends his time in sports and calls it recreation is like him whose garment is all made of fringes, and his meat nothing but sauces; they are healthless, chargeable, and useless.

Set apart some portions of every day for more solemn devotion and religious employment, which be severe in observing. And if variety of employment, or prudent affairs, or civil society, press upon you, yet so order thy rule that the necessary parts of it be not omitted.

When the clock strikes, or however else you shall measure the day, it is good to say a short ejaculation every hour, that the parts and returns of devotion may be the measure of your time. And do so also in all the breaches of thy sleep; that those spaces which have in them no direct business of the world may be filled with religion.

We shall be much assisted if, before we sleep, every night we examine the actions of the past day with a particular scrutiny, if there have been any accident extraordinary; as long discourse, a feast, much business, variety of company. If nothing but common hath happened, the less examination will suffice.

Let all these things be done prudently and moderately, not with scruple and vexation. For these are good advantages, but the particulars are not divine commandments, and therefore are to be used as shall be found expedient to everyone's condition. For provided that our duty be secured, for the degrees and for the instruments every man is permitted to himself and the conduct of such who shall be appointed to him.

THE PRACTICE OF THE PRESENCE OF GOD.

If men would always actually consider and really esteem this truth, that God is the great eye of the world, always watching over our actions, and an ever-open ear to all our words, and an unwearied arm ever lifted up to crush a sinner into ruin, it would be the readiest way in the world to make sin to cease from amongst the children of men, and for men to approach to the blessed estate of the saints in heaven, who cannot sin, for they always walk in the presence and behold the face of God. If you will sin, go where God cannot see, for nowhere else can you be safe.

Let everything you see represent to your spirit the presence, the excellency, and the power of God. And let your conversation with the creatures lead you unto the Creator. For so shall your actions be done more frequently with an actual eye to God's presence by your often seeing him in the glass of the creation. In the face of the sun you may see God's beauty; in the fire you may feel his heat warming; in the water, his gentleness to refresh you; he it is that comforts your spirits when you have taken cordials.

In your retirement make frequent colloquies, or short discoursings, between God and thine own soul. Every act of complaint or thanksgiving, every petition and every return of the heart in these intercourses, is a going to God and appearing in his presence, and a representing him present to thy spirit and to thy necessity. And this was long since, by a spiritual person, called "a building to God a chapel in our heart." It reconciles Martha's employment with Mary's devotion, charity, and religion. For thus in the midst of the works of your trade you may retire into your chapel, your heart, and converse with God by frequent addresses and returns.

Let us remember that God is in us, and that we are in him; we are his workmanship, let us not deface it; we are in his presence, let us not pollute it by unholy actions.

God is in every place; suppose it, therefore, to be a church. And that decency of deportment and piety of carriage which you are taught by religion or by custom, or by civility and public manners, to use in churches, the same use in all places.

God is in every creature. Be cruel toward none, neither abuse any by intemperance. Remember that the creatures, and every member of thy body, is one of the lesser cabinets and receptacles of God.

He walks as in the presence of God that converses with him in frequent prayer and frequent communion; that runs to him in all his necessities; that asks counsel of him in all his doubtings; that opens all his wants to him; that weeps before him for his sins; that asks remedy and support for his weakness; that fears him as a judge, reverences him as a lord, obeys him as a father, and loves him as a patron.

HUMILITY.

Humility is the great ornament and jewel of Christian religion, that whereby it is distinguished from all the wisdom of the world; it not having been taught by the wise men of the Gentiles, but first put into a disciple and made part of a religion by our Lord Jesus Christ.

Think not thyself better for anything that happens to thee from without. Whatsoever other difference there is between thee and thy neighbor, if it be bad it is thine own, but thou hast no reason to boast of thy misery and shame; if it be good, thou hast received it from God, and then thou art more obliged to pay duty and tribute, use and principal, to him; and it were a strange folly for a man to be proud of being more in debt than another.

Never speak anything directly tending to thy praise or glory; that is, with a purpose to be commended, and for no other end. If other ends be mingled with thy honor, as if the glory of God, or charity, or necessity, or anything of prudence be thy end, you are not tied to omit your discourse or your design that you may avoid praise, but pursue your end, though praise come along in the company. Only let not praise be the design.

When thou hast said or done a thing for which thou receivest praise or estimation, take it indifferently, and return it to God, reflecting upon him as the giver of the gift, or the blesser of the action, or the aid of the design; and give God thanks for making thee an instrument of his glory, for the benefit of others.

Secure a good name to thyself by living virtuously and humbly; but let this good name be nursed abroad, and never be brought home to look upon it. Let others use it for their own advantage; let them speak of it, if they please; but do not thou at all use it but as an instrument to do God glory and thy neighbor more advantage. Let thy face, like Moses', shine to others, but make no looking-glasses for thyself.

Take no content in praise when it is offered thee, but let thy rejoicing in God's gift be alloyed with fear lest this good bring thee to evil. Use the praise as you use your pleasure in eating and drinking. If it comes make it do drudgery, let it serve other ends, and minister to necessities, and to caution, lest by pride you lose your just praise which you have deserved; or else, by being praised unjustly, you receive shame unto yourself with God and wise men.

Use no stratagems and devices to get praise. Some use to inquire into the faults of their own actions or discourses on purpose to hear that it was well done or spoken and without fault. Others bring the matter into talk, or thrust themselves into company, and intimate and give occasion to be thought or spoken of. These men make a bait to persuade themselves to swallow the hook, till by drinking the waters of vanity they swell and burst.

Suffer others to be praised in thy presence, and entertain their good and glory with delight; but at no hand disparage them, or lessen the report, or make an objection; and think not the advancement of thy brother is a lessening of thy worth. Be content that he should be employed, and thou laid by as unprofitable; his sentence approved, thine rejected; he be preferred, and thou fixed in a low employment.

Never compare thyself with others, unless it be to advance them and to depress thyself. To which purpose we must be sure in some sense or other to think ourselves the worst in every company where we come. One is more learned than I am, another is more prudent, a third more honorable, a fourth more chaste, or he is more charitable, or less proud. For the humble man observes their good, and reflects only upon his own vileness; or considers the many evils of himself certainly known to himself, and the ill of others but by uncertain report. Or he considers that the evils done by another are out of much infirmity or ignorance, but his own sins are against a clearer light; and if the other had so great helps, he would have done more good and less evil. Or he remembers that his old sins before his conversion were greater in the nature of the thing, or in certain circumstances, than the sins of other men.

Make no reflex acts upon thine own humility, nor

upon any other grace with which God hath enriched thy soul. Spiritual pride is very dangerous, not only by reason it spoils so many graces by which we drew nigh unto the kingdom of God, but also because it so frequently creeps in upon the spirit of holy persons.

Remember that the blessed Saviour of the world hath done more to prescribe and transmit and secure this grace than any other; his whole life being a great, continued example of humility.

Drive away all flatterers from thy company, and at no hand endure them; for he that endures himself so to be abused by another is not only a fool for entertaining the mockery, but loves to have his own opinion of himself to be heightened and cherished.

Never change thy employment for the sudden coming of another to thee; but if modesty permits, or discretion, appear to him that visits thee the same that thou wert to God and thyself in thy privacy. But if thou wert walking or sleeping, or in any other innocent employment or retirement, snatch not up a book to seem studious, nor fall on thy knees to seem devout, nor alter anything to make him believe thee better employed than thou wert.

The humble man does not pertinaciously pursue the choice of his own will, but in all things lets God choose for him, and his superiors in those things which concern them. He does not murmur against commands. He is meek and indifferent in all accidents and chances. He patiently bears injuries. He is always unsatisfied in his own conduct, resolutions, and counsels. He is a great lover of good men, and a praiser of wise men, and a censurer of no man. He fears when he hears himself commended, lest God make another judgment concerning his actions than men do. He loves to sit down in private and, if he may, refuses the temptation of offices and new honors. He mends his fault, and gives thanks, when he is admonished.

CONTENTEDNESS.

Here is the wisdom of the contented man, to let God choose for him. For when we have given up our wills to him, and stand in that station of the battle where our great general hath placed us, our spirits must needs rest, while our conditions have, for their security, the power, the wisdom, and the charity of God. For no man is poor that does not think himself so; but if, in a full fortune, with impatience he desires more, he proclaims his wants and his beggarly condition.

Contentedness in all accidents brings great peace of spirit, and is the great and only instrument of temporal felicity. It removes the sting from the accident, and makes a man not to depend upon chance and the uncertain dispositions of men for his well-being, but only on God and his own spirit. We ourselves make our fortunes good or bad; and when God lets loose a tyrant upon us, or a sickness, or scorn, or a lessened fortune, if we fear to die, or know not to be patient, or are proud, or covetous, then the calamity sits heavy on us. But if we know how to manage a noble principle, and fear not death so much as a dishonest action, and think impatience a worse evil than a fever, and pride to be the biggest disgrace, and poverty to be infinitely desirable before the torments of covetousness—then we, who now think vice to be so easy, and make it so familiar, and think the cure so impossible, shall quickly be of another mind, and reckon these accidents amongst things eligible.

But no man can be happy that hath great hopes and great fears of things without, and events depending upon other men, or upon the chances of fortune. He that suffers a transporting passion concerning things within the power of others is free from sorrow and amazement no longer than his enemy shall give him leave; and it is ten to one but he shall be smitten then and there where it shall most trouble him.

When anything happens to our displeasure let us endeavor to take off its trouble by turning it into spiritual or artificial advantage, and handle it on that side in which it may be useful to the designs of reason. For there is nothing but hath a double handle, or at least we have two hands to apprehend it. If thou fallest from thy employment in public take sanctuary in an honest retirement, being indifferent to thy gain abroad or thy safety at home. If a calamity does any good to our souls it hath made more than sufficient recompense for all the temporal affliction.

Never compare thy condition with those above thee; but, to secure thy content, look upon those thousands with whom thou wouldst not, for any interest, change thy fortune and condition. There is no wise or good man that would change persons or conditions entirely with any man in the world. It may be he would have one man's wealth added to himself, or the power of a second, or the learning of a third; but still he would receive these into his own person, because he loves that best, and therefore esteems it best, and therefore overvalues all that which he is before all that which any other man in the world can be. Either change all or none. Cease to love yourself best, or be content with that portion of being and blessing for which you love yourself so well.

It conduces much to our content if we pass by these things which happen to our trouble, and consider that which is pleasing and prosperous, that, by the representation of the better, the worse may be blotted out. Or else reckon the blessings which already you have received, and therefore be pleased, in the change and variety of affairs, to receive evil from the hand of God as well as good. Or else please thyself with hopes of the future. Harvest will come, and then every farmer is rich, at least for a month or two. It may be thou art entered into the cloud which will bring a gentle shower to refresh thy sorrows. When a sadness lies heavy upon thee remember that thou art a Christian designed to the inheritance of Jesus; and what dost thou think concerning thy great fortune, thy lot, and portion of eternity?

These arts of looking backward and forward are more than enough to support the spirit of a Christian; there is no man but hath blessings enough in present possession to outweigh the evils of great affliction. If you miss an office for which you stood candidate, then, besides that you are quit of the cares and the envy of it, you still have all those excellencies which rendered you capable to receive it, and they are better than the best office in the commonwealth. Or I am fallen into the hands of publicans and sequestrators, and they have taken all from me. What now? Let me look about me. They have left me the sun and moon, fire and water, a loving wife, and many friends to pity me, and some to relieve me, and I can still discourse; and, unless I list, they have not taken away my merry countenance, and my cheerful spirit, and a good conscience; they still have left me the providence of God, and all the promises of the Gospel, and my religion, and my hopes of heaven, and my charity to them too; and still I sleep and digest, I eat and drink, I read and meditate, I can walk in my neighbor's pleasant fields, and see the varieties of natural beauties, and delight in all that in which God delights, that is, in virtue and wisdom, in the whole creation, and in God himself. And he that hath so many causes of joy, and so great, is very much in love with sorrow and peevishness who loses all these pleasures and chooses to sit down upon his little handful of thorns.

Enjoy the present, whatsoever it be, and be not solicitous for the future. "Sufficient to the day," said Christ, "is the evil thereof;" sufficient, but not intolerable. But if we look abroad and bring into one day's thoughts the evil of many, certain and uncertain, what will be and what will never be, our load will be as intolerable as it is unreasonable.

Let us prepare our minds against changes, always expecting them, that we be not surprised when they come; for nothing is so great an enemy to tranquillity and a contented spirit as the amazement and confusions of unreadiness and inconsideration.

Let us often frame to ourselves, and represent to our considerations, the images of those blessings we have, just as we usually understand them when we want them. Consider how desirable health is to a sick man, or liberty to a prisoner; and if but a fit of the toothache seizes us with violence all those troubles which in our health afflicted us disband instantly and seem inconsiderable.

If you will secure a contented spirit, you must measure your desires by your fortune and condition, not your fortunes by your desires. That is, be governed by your needs, not by your fancy; by nature, not by evil customs and ambitious principles.

Consider that the universal providence of God hath so ordered it that the good things of nature and fortune are divided, that we may know how to bear our own and relieve each other's wants and imperfections. It is not for a man, but for a God, to have all excellencies and all felicities.

Consider how many excellent personages in all ages have suffered as great or greater calamities than this which now tempts thee to impatience. It were a strange pride to expect to be more gently treated by the divine Providence than the best and wisest men, than apostles and saints, nay, the Son of the eternal God, the heir of both the worlds.

There are many accidents which are esteemed great calamities, and yet we have reason enough to bear them well and unconcernedly; for they neither touch our bodies nor our souls: our health and our virtue remain entire, our life and our reputation. Inquire what you are the worse, either in your soul or in your body, for what hath happened; for upon this very stock many evils will disappear, since the body and the soul make up the whole man.

Consider that sad accidents and a state of affliction is a school of virtue; it reduces our spirits to soberness, and our counsels to moderation; it corrects levity, and interrupts the confidence of sinning.

Consider that afflictions are oftentimes the occasions of great temporal advantages; and we must not look upon them as they sit down heavily upon us, but as they serve some of God's ends, and the purposes of universal Providence. For God esteems it one of his glories that he brings good out of evil; and therefore it were but reason we should trust God to govern his own world as he pleases, and that we should patiently wait till the change cometh or the reason be discovered. To which also may be added that the great evils which happen to the best and wisest men are one of the great arguments upon the strength of which we can expect felicity to our souls and the joys of another world. And certainly they are then very tolerable and eligible when, with so great advantages, they minister to the faith and hope of a Christian.

LOVE TO GOD.

Love does all things which may please the beloved person; it performs all his commandments. Love is obedient. It does all the intimations and secret significations of his pleasure whom we love. Great love is pliant and inquisitive in the instances of its expression.

Love gives away all things, that so he may advance the interest of the beloved person. He never loved God that will quit anything of his religion to save his money. Love is always liberal and communicative.

It suffers all things that are imposed by its beloved, or that can happen for his sake, or that intervene in his service, cheerfully, sweetly, willingly; expecting that God should turn them into good, and instruments of felicity.

Love is also impatient of anything that may displease the beloved person; hating all sin as the enemy of its friend; for love contracts all the same relations, and marries the same friendships and the same hatreds. And all affection to a sin is perfectly inconsistent with the love of God.

Love endeavors forever to be present, to converse with, to enjoy, to be united with its object; loves to be talking of him, reciting his praises, telling his stories, repeating his words, imitating his gestures, transcribing his copy in everything; and every degree of union and every degree of likeness is a degree of love; and it can endure anything but the displeasure and the absence of its beloved.

He that loves God is not displeased at those accidents which God chooses; nor murmurs at those changes which he makes in his family; nor envies at those gifts he bestows: but chooses as he likes, and is ruled by his judgment, and is perfectly of his persuasion; loving to learn where God is the teacher, and being content to be ignorant or silent where he is not pleased to open himself.

Love is curious of little things, or circumstances and measures, and little accidents; not allowing to itself any infirmity which it strives not to master, aiming at what it cannot yet reach, desiring to be of an angelical purity, and of a perfect innocence, and of a seraphical fervor, and fears every image of offense; is as much afflicted at an idle word as some at an act of adultery, and will not allow to itself so much anger as will disturb a child, nor endure the impurity of a dream. And this is the curiosity and niceness of Divine love; this is the fear of God, and is the daughter and production of love.

FÉNELON.

FOR two hundred years Fénelon has stood among the choicest few of those universally esteemed to be best qualified as religious guides. He belongs to no age and to no Church, but to all. He exemplified so sweetly in his life what he preached, and preached so eloquently what he lived, that few indeed have ever been found to equal him as an authority in spiritual things. He not only had a heart filled with the love of God and glowing with pure devotion, but also a mind capable of the closest analysis and the keenest discrimination. He was not only a saint, but also a scholar and a genius. Such combinations are very rare. His thirst for perfection has probably never been surpassed. He follows self-love into its minutest workings, exposes all its subtleties, gives it no quarter, insists that it shall be destroyed root and branch.

Fénelon sprang from one of the most illustrious families of France, his full name being François de Salignac de la Mothe Fénelon; his birthday was August 6, 1651. His constitution was delicate, his natural disposition extremely amiable, his education conducted mainly at the College of Cahors and in Paris at the College du Plessis. He began to preach, attracting much attention, at the age of fifteen. Twice he seriously contemplated giving himself to the work of foreign missions, but was prevented from carrying out his design. He was for some years preceptor to the Duke of Burgundy, the son of the dauphin, and at the age of forty-three he became Archbishop of Cambray. He was everywhere known as "the good archbishop." No act of kindness was so great as to overtask him or so small as to escape his notice. His purity and gentleness of spirit subdued his enemies. The fullness of his love to all made it easy for him to extend forgiveness, and the freedom of his mind from vanity, as well as the exquisite courtesy of his manner, put everyone at ease in his presence. His sermons were always the outpourings of his heart. So extensive had been his charities, and yet so well balanced his worldly affairs, that he died without money and without a debt. He departed this life January 7, 1715, exhibiting in his last illness the same sweetness of temper, composure of mind, love for his fellow-men, and confidence in God which distinguished all his days. He had the spirit of the Saviour in an extremely high degree, and came as near, perhaps, as any human being has done to losing his own will in the will divine.

He was a voluminous writer. The most complete edition of his works, published at Versailles about seventy years ago, is comprised in thirty-four octavo volumes. Many of his writings have been translated into English, and various selections from them have been published from time to time. The extracts that follow are mostly taken from his *Spiritual Letters* and his *Christian Counsel on Divers Matters Pertaining to the Inner Life.*

DAILY FAULTS.

Little faults become great in our eyes in proportion as the pure light of God increases in us; just as the sun in rising reveals the true dimensions of objects which were dimly and confusedly discovered during the night. Be sure that with the increase of the inward light the imperfections which you have hitherto seen will be beheld as far greater and more deadly in their foundations than you now conceive them, and that you will witness, in addition, the development of a crowd of others of the existence of which you have not now the slightest suspicion. You will find the weaknesses necessary to deprive you of all confidence in your own strength; but this discovery, far from discouraging, will but serve to destroy your self-reliance, and to raze to the ground the edifice of pride.

Our faults, even those most difficult to bear, will all be of service to us if we make use of them for our humiliation without relaxing our efforts to correct them. We must bear with ourselves without either flattery or discouragement, a mean seldom attained. Utter despair of ourselves, in consequence of a conviction of our helplessness, and unbounded confidence in God, are the true foundations of the spiritual edifice.

Faults of haste and frailty are nothing in comparison with those where we shut our eyes to the voice of the Holy Spirit beginning to speak in the depths of the heart.

Discouragement is not a fruit of humility, but of pride; nothing can be worse. Suppose we have stumbled, or even fallen, let us rise and run again; all our falls are useful if they strip us of a disastrous confidence in ourselves, while they do not take away a humble and salutary trust in God.

Carefully purify your conscience from daily faults; suffer no sin to dwell in your heart; small as it may seem, it obscures the light of grace, weighs down the soul, and hinders that constant communion with Jesus Christ which it should be your pleasure to cultivate; you will become lukewarm, forget God, and find yourself growing in attachment to the creature. The great point is never to act in opposition to the inward light, and to be willing to go as far as God would have us.

St. Francis of Sales says that great virtues and fidelity in small things are like sugar and salt; sugar is more delicious but of less frequent use, while salt enters into every article of our food. Small occasions are unforeseen; they recur every moment, and place us incessantly in conflict with our pride, our sloth, our self-esteem, and our passions; they are calculated thoroughly to subdue our wills and leave us no retreat. If we are faithful in them nature will have no time to breathe, and must die to all her inclinations. It would please us much better to make some great sacrifices, however painful and violent, on condition of obtaining liberty to follow our own pleasure, and retain our old habits in little things. But it is only by this fidelity in small matters that the grace of true love is sustained and distinguished from the transitory excitements of nature.

God does not so much regard our actions as the motive of love from which they spring, and the pliability of our wills to his. Men judge our deeds by their outward appearance; with God that which is most dazzling in the eyes of man is of no account. What he desires is a pure intention, a will ready for anything and ever pliable in his hands, and an honest abandonment of self; and all this can be much more frequently manifested on small than on extraordinary occasions; there will also be much less danger from pride, and the trial will be far more searching. Indeed, it sometimes happens that we find it harder to part with a trifle than with an important interest; it may be more of a cross to abandon a vain amusement than to bestow a large sum in charity.

The greatest danger of all consists in this, that by neglecting small matters the soul becomes accustomed to unfaithfulness. We grieve the Holy Spirit, we return to ourselves, we think it a little thing to be wanting toward God. On the other hand, true love can see nothing small; everything that can either please or displease God seems to be great; not that true love disturbs the soul with scruples, but it puts no limits to its faithfulness. It acts simply with God; and as it does not concern itself about those things which God does not require from it, so it never hesitates an instant about those which he does, be they great or small.

Thus it is not by incessant care that we become faithful and exact in the smallest things, but simply by a love which is free from the reflections and fears of restless and scrupulous souls. We are, as it were. drawn along by the love of God; we have no desire to do anything but what we do, and no will in respect to anything which we do not do. The soul enjoys perfect peace in God.

NOT PERFECT IN A MOMENT.

Neither in his gracious nor providential dealings does God work a miracle lightly. It would be as great a wonder to see a person full of self become in a moment dead to all self-interest, and all sensitiveness, as it would be to see a slumbering infant wake in the morning a fully developed man. God works in a mysterious way in grace as well as in nature, concealing his operations under an imperceptible succession of events, and thus keeps us always in the darkness of faith.

He makes use of the inconstancy and ingratitude of the creature, and of the disappointments and surfeits which accompany prosperity, to detach us from them both; he frees us from self by rendering to us our weaknesses, and our corruptions, in a multitude of backslidings. All this dealing appears perfectly natural, and it is by this succession of natural means that we are burnt as by a slow fire. We should like to be consumed at once by the flames of pure love, but such an end would scarce cost us anything; it is only an excessive self-love that desires thus to become perfect in a moment and at so cheap a rate.

We cling to an infinity of things which we should never suspect; we only feel that they are a part of us when they are snatched away, as I am only conscious that I have hairs when they are pulled from my head. God develops to us, little by little, what is within us, of which we are, until then, entirely ignorant, and we are astonished at discovering in our very virtues defects of which we should never have believed ourselves capable.

God spares us by discovering our weakness to us just in proportion as our strength to support the view of it increases. We discover our imperfections one by one, as we are able to cure them. Without this merciful preparation, that adapts our strength to the light within, we should be in despair.

To the sincere desire to do the will of God we must add a cheerful spirit that is not overcome when it has failed, but tries again and again to do better; hoping always to the very end to be able to do it; bearing with its own involuntary weakness as God bears with it; waiting with patience for the moment when it shall be delivered from it; going straight on in singleness of heart according to the strength that it can command; losing no time by looking back, nor making useless reflections upon its falls, which can only embarrass and retard its progress. The first sight of our little failings should humble us, but then we must press on; not judging ourselves with a Judaical rigor, not regarding God as a spy watching for our least offense, or as an enemy who places snares in our path, but as a Father who loves and wishes to save us; trusting in his goodness, invoking his blessing, and doubting all other support. This is true liberty.

One of the principles in the doctrines of holy living is that we should not be premature in drawing the conclusion that the process of inward crucifixion is complete, and that our abandonment to God is without any reservation whatever. The act of consecration, which is a sort of incipient step, may be sincere; but the reality of the consecration in the full extent to which we suppose it to exist, and which may properly be described as entire self-renunciation, can be known only when God has applied the appropriate tests. The trial will show whether we are wholly the Lord's. Those who prematurely draw the conclusion that they are so expose themselves to great illusion and injury.

EASY WAYS OF DIVINE LOVE.

Christian perfection is not that rigorous, tedious, cramping thing that many imagine. It demands only an entire surrender of everything to God from the depths of the soul, and the moment this takes place whatever is done for him becomes easy. They who are God's without reserve are in every state content; for they will only what he wills, and desire to do for him whatever he desires them to do; they strip themselves of everything, and in this nakedness find all things restored a hundredfold. Peace of conscience, liberty of spirit, the sweet abandonment of themselves and theirs into the hand of God, the joy of perceiving the light always increasing in their hearts, and finally the freedom of their souls from the bondage of the fears and desires of this world—these things constitute that return of happiness which the true children of God receive a hundredfold in the midst of their crosses, while they remain faithful.

What God requires of us is a will which is no longer divided between him and any creature; a simple, pliable state of will which desires what he desires, rejects nothing but what he rejects, and wills without reserve what he wills, and under no pretext wills what he does not. In this state of mind all things are proper for us; our amusements, even, are acceptable in his sight.

Blessed is he who thus gives himself to God! He is delivered from his passions, from the opinions of men, from their malice, from the tyranny of their maxims, from their cold and miserable raillery, from the misfortunes which the world attributes to chance, from the infidelity and fickleness of friends, from the artifices and snares of enemies, from the wretchedness and shortness of life, from the horrors of an ungodly death, from the cruel remorse that follows sinful pleasures, and, finally, from the everlasting condemnation of God.

Happy those who throw themselves, as it were, headlong, and with their eyes shut, into the arms of "the Father of mercies and the God of all comfort." Their whole desire then is to know what is the will of God respecting them; and they fear nothing so much as not perceiving the whole of his requirements. So soon as they behold a new light in his law they are transported with joy, like a miser at the finding of a treasure.

No matter what cross may overwhelm the true child of God, he wills everything that happens, and would not have anything removed which his Father appoints; the more he loves God, the more is he filled with content; and the most stringent perfection, far from being a burden, only renders his yoke the lighter.

THE DIVINE PRESENCE.

The true source of all our perfection is contained in the command of God to Abraham, "Walk before me, and be thou perfect" (Gen. xvii, 1).

The presence of God calms the soul, and gives it quiet and repose even during the day and in the midst of occupation; but we must be given up to God without reserve.

Whenever we perceive within us anxious desires for anything, whatever it may be, and find that nature is hurrying us with too much haste to do what is to be done, whether it be to see something, say something, or to do something, let us stop short and repress the precipitancy of our thoughts and the agitations of our actions; for God has said that his Spirit does not dwell in disquiet.

An excellent means of preserving our interior solitude and liberty of soul is to make it a rule to put an end, at the close of every action, to all reflections upon it, all reflex acts of self-love, whether of a vain joy or sorrow.

Let us be accustomed to recollect ourselves during the day and in the midst of our occupations by a simple view of God. Let us silence by that means all the movements of our hearts, when they appear in the least agitated. Let us separate ourselves from all that does not come from God. Let us suppress our superfluous thoughts and reveries. Let us utter no useless word. Let us seek God within us, and we shall find him without fail, and with him joy and peace.

While outwardly busy let us be more occupied with God than with everything else. To be rightly engaged we must be in his presence and employed for him. At the sight of the majesty of God our interior ought to become calm and remain tranquil. Once a single word of the Saviour suddenly calmed a furiously agitated sea; one look of his at us, and of ours toward him, ought always to perform the same miracle within us.

We must not wait for a leisure hour when we can bar our doors; the moment that is employed in regretting that we have no opportunity to be recollected might be profitably spent in recollection. Let us turn our hearts toward God in a simple, familiar spirit, full of confidence in him. The most interrupted moments, even while eating, or listening to others, are valuable. Tiresome and idle talk in our presence, instead of annoying, will afford us the delight of employing the interval in seeking God. Thus all things work together for good to them that love God.

Let us be careful not to suffer ourselves to be overwhelmed by the multiplicity of our exterior occupations, be they what they may. Let us endeavor to commence every enterprise with a pure view to the glory of God, continue it without distraction, and finish it without impatience. The intervals of relaxation and amusement are the most dangerous for us and perhaps the most useful for others; we must then be on our guard that we be as faithful as possible to the presence of God. We can never employ our leisure hours better than in refreshing our spiritual strength by a secret and intimate communion with God. Prayer is so necessary, and the source of so many blessings, that he who has discovered the treasure cannot be prevented from having recourse to it whenever he has an opportunity.

TRUE PRAYER.

True prayer is only another name for the love of God. To pray is to desire—but to desire what God would have us desire. He who asks what he does not from the bottom of his heart desire is mistaken in thinking that he prays. O how few there are who pray! for how few are they who desire what is truly good. Crosses, external and internal humiliation, renouncement of our own wills, the death of self and the establishment of God's throne upon the ruins of self-love—these are indeed good; not to desire these is not to pray; to desire them seriously, soberly, constantly, and with reference to all the details of life—this is true prayer. Alas! how many souls full of self, and of an imaginary desire for perfection in the midst of hosts of voluntary imperfections, have never yet uttered this true prayer of the heart! It is in reference to this that St. Augustine says: "He that loveth little prayeth little; he that loveth much prayeth much."

Our intercourse with God resembles that with a friend: at first there are a thousand things to be told, and as many to be asked; but after a time these diminish, while the pleasure of being together does not. Everything has been said, but the satisfaction of seeing each other, of feeling that one is near the other, of reposing in the enjoyment of a pure and sweet friendship, can be felt without conversation; the silence is eloquent and mutually understood. Each feels that the other is in perfect sympathy with him, and that their two hearts are incessantly poured out into the other, and constitute but one.

Those who have stations of importance to fill have generally so many indispensable duties to perform that, without the greatest care in the management of their time, none will be left to be alone with God. If they have ever so little inclination for dissipation the hours that belong to God and their neighbor disappear altogether. We must be firm in observing our rules. This strictness seems excessive, but without it everything falls into confusion; we become dissipated, relaxed, and lose strength; we insensibly separate from God, surrender ourselves to all our pleasures, and only then begin to perceive that we have wandered where it is almost hopeless to think of endeavoring to return.

The Christian life is a long and continual tendency of our hearts toward that eternal goodness which we desire upon earth. All our happiness consists in thirsting for it. Now this thirst is prayer. Ever desire to approach your Creator and you will never cease to pray.

The best of all prayers is to act with a pure intention and with a continual reference to the will of God. Unhappy are they whose prayers do not render them more humble, more submissive, more vigilant over their faults, and more willing to live in obscurity. The coldness of our love is the silence of our hearts toward God. Without this we may pronounce prayers, but we do not pray; for what shall lead us to meditate upon the laws of God if it be not the love of him who has made these laws?

THE HUMAN WILL.

True virtue and pure love reside in the will alone. The question is not, What is the state of our feelings? but, What is the condition of our will? Let us will to have whatever we have, and not to have whatever we have not. We would not even be delivered from our sufferings, for it is God's place to apportion to us our crosses and our joys. In the midst of affliction we rejoice, as did the apostle; but it is not joy of the feelings, but of the will. The wicked are wretched in the midst of their pleasures, because they are never content with their state; they are always desiring to remove some thorn, or to add some flower to their present condition. The faithful soul, on the other hand, has a will which is perfectly free; it accepts, without questioning, whatever bitter blessings God develops, wills them, loves them, and embraces them; it would not be freed from them if it could be accomplished by a simple wish; for such a wish would be an act originating in self and contrary to its abandonment to Providence, and it is desirous that this abandonment should be absolutely perfect.

If there be anything capable of setting a soul in a large place it is this absolute abandonment to God. If there be anything that can render the soul calm, dissipate its scruples and dispel its fears, sweeten its sufferings by the anointing of love, impart strength to it in all its actions, and spread abroad the joy of the Holy Spirit in its countenance and words, it is this simple, free, and childlike repose in the arms of God.

The important question is, not how much you enjoy religion, but whether you will whatever God wills.

The essence of virtue consists in the attitude of the will. That kingdom of God which is within us consists in our willing whatever God wills, always, in everything, and without reservation; and thus his kingdom comes; for his will is then done as it is in heaven, since we will nothing but what is dictated by his sovereign pleasure. Thus nothing can ever come to pass against our wishes; for nothing can happen contrary to the will of God, and we find in his good pleasure an inexhaustible source of peace and consolation.

The interior life is the beginning of the blessed peace of the saints, who eternally cry, Amen, Amen. We adore, we praise, we bless God in everything; we see him incessantly, and in all things his paternal hand is the sole object of our contemplation. There are no longer any evils; for even the most terrible that can come upon us work together for our good. Can the suffering that God designs to purify us and make us worthy of himself be called an evil?

Let us cast all our cares then into the bosom of so good a Father, and suffer him to do as he pleases. Let us be content to adopt his will in all points, and to abandon our own absolutely and forever. How can we retain anything of our own when we do not even belong to ourselves? The only thing that really belongs to us is our will, and it is of this, therefore, that God is especially jealous, for he gave it to us not that we should retain it, but that we should return it to him, whole as we received it, and without the slightest reservation. If the least desire remains, or the smallest hesitation, it is robbing God, contrary to the order of creation; for all things come from him, and to him they are all due. Alas! how many souls there are full of self, and desirous of doing good and serving God, but in such a way as to suit themselves; who desire to impose rules upon God as to his manner of drawing them to himself. They want to serve and possess him, but they are not willing to abandon themselves to him and be possessed by him.

To desire to serve God in one place rather than in another, in this way rather than in that—is not this desiring to serve him in our own way rather than in his? But to be equally ready for all things, to will everything and nothing, to leave ourselves in his hands like a toy in the hands of a child, to set no bounds to our abandonment inasmuch as the perfect reign of God cannot abide them—this is really denying ourselves; this is treating him like a God and ourselves like creatures made solely for his use.

The peace of the soul consists in an absolute resignation to the will of God. The pain we suffer from so many occurrences arises from the fact that we are not entirely abandoned to God in everything that happens. Let us put all things, then, into his hands, and offer them to him in our hearts, as a sacrifice beforehand. From the moment that you cease to desire anything according to your own judgment, and begin to will everything just as God wills it, you will be free from your former tormenting reflections and anxieties about your own concerns; you will no longer have anything to conceal or take care of.

CONTINUAL CROSSES.

In regard to austerities everyone must regard his attraction, his state, his need, and his temperament. A simple mortification, consisting in nothing more than an unshaken fidelity in providential crosses, is often far more valuable than severe austerities which render the life more marked, and tempt to a vain self-complacency. Whoever will refuse nothing which comes in the order of God, and seek nothing out of that order, need never fear to finish his day's work without partaking of the cross of Jesus Christ. There is an indispensable Providence for crosses as well as for the necessaries of life; they are a part of our daily bread; God never will suffer it to fail. It is sometimes a very useful mortification to certain fervent souls to give up their own plans of mortification and adopt with cheerfulness those which are momentarily revealed in the order of God. When a soul is not faithful in providential mortifications there is reason to fear some illusion in those which are sought through the fervor of devotion; such warmth is often deceitful, and it seems to me that a soul in this case would do well to examine its faithfulness under the daily crosses allotted by Providence.

The crosses which originate with ourselves are not as efficient in eradicating self-love as those which come in the daily allotments of God. These latter contribute no aliment for the nourishment of our own wills, and as they proceed immediately from a merciful Providence they are accompanied by grace sufficient for all our needs. We have nothing to do, then, but to surrender ourselves to God each day, without looking further; he will carry us in his arms as a tender mother bears her child.

The best rule we can ever adopt is to receive equally, and with the same submission, everything that God sends us during the day, both within and without. Without, there are things disagreeable that must be met with courage, and things pleasant that must not be suffered to arrest our affections. They must be received because God sends them, and not because they are agreeable to our own feelings; they are to be used, like any other medicine, without self-complacency, without attachment to them, and without appropriation. We must accept them, but not hold on to them; so that when God sees fit to withdraw them we may neither be dejected nor discouraged. We must count less upon sensible delights, and the measures of wisdom which we devise for our own perfection, than upon simplicity, lowliness, renunciation of our own efforts, and perfect pliability to all the designs of grace.

WHAT IS MEANT BY RENOUNCING ALL?

We must not only renounce evil, but also good things; for Jesus has said, "Whosoever he be of you that renounceth not all that he hath, he cannot be my disciple" (Luke xiv, 33). The abandonment of evil things consists in refusing them with horror; of good things, in using them with moderation for our necessities, continually studying to retrench all those imaginary wants with which greedy nature would flatter herself. We are moderately, and without inordinate emotion, to do what is in our power to retain goods and honors in order to make a sober use of them, without desiring to enjoy them, or placing our hearts upon them.

The Christian must abandon everything that he has, however innocent; for if he do not renounce it it ceases to be innocent. He must abandon those things which it is his duty to guard with the greatest possible care, such as the good of his family, or his own reputation, for he must have his heart on none of these things; he must be ready to give them all up whenever it is the will of Providence to deprive him of them.

He must give up those whom he loves best, and whom it is his duty to love; and his renouncement of them consists in this, that he is to love them for God only; to make use of the consolation of their friendship soberly, and for the supply of his wants; to be ready to part with them whenever God wills it, and never to seek in them the true repose of his heart. It is thus that we use the world and the creature as not abusing them. We do not desire to take pleasure in them; we only use what God gives us, what he wills that we should love, and what we accept with the reserve of a heart receiving it only for necessity's sake and keeping itself for a more worthy object. It is in this sense that Christ would have us leave father and mother, brothers and sisters and friends, and that he is come to bring a sword upon earth.

Having abandoned everything exterior, it remains to complete the sacrifice by renouncing everything interior, including self. You must renounce all satisfaction, and all natural complacency in your own wisdom and virtue. Remember, the purer and more excellent the gifts of God the more jealous he is of them. He would have us attached to nothing but himself, and to regard his gifts, however excellent, as only the means of uniting us more easily and intimately to him. Whoever contemplates the grace of God with a satisfaction and sort of pleasure of ownership turns it into poison.

Live, as it were, on trust; all that is in you, and all that you are, is only loaned you; make use of it according to the will of Him who lends it, but never regard it for a moment as your own. Herein consists true self-abandonment; it is this spirit of self-divesting, this use of ourselves and of ours with a single eye to the movements of God, who alone is the true proprietor of his creatures. You may be exercised in self-renunciation in every event of every day.

Happy is he who never hesitates; who fears only that he follows with too little readiness; who would rather do too much against self than too little! Blessed is he who, when asked for a sample, boldly presents his entire stock and suffers God to cut from the whole cloth! Happy he who, esteeming himself as nothing, puts God to no necessity of sparing him! Thrice happy he whom all this does not affright! It is thought that this state is a painful one; it is a mistake: here is peace and liberty; here the heart, detached from everything, is immeasurably enlarged, so as to become illimitable; nothing cramps it; and in accordance with the promise it becomes, in a certain sense, one with God himself.

True progress does not consist in a multitude of views, nor in austerities, trouble, and strife; it is simply willing nothing and everything, without reservation and without choice, cheerfully performing each day's journey as Providence appoints it for us; seeking nothing, refusing nothing; finding everything in the present moment, and suffering God, who does everything, to do his pleasure in and by us without the slightest resistance. O, how happy is he who has attained to this state! and how full of good things is his soul when it appears emptied of everything!

HOW TO WATCH.

The soul which God truly leads by the hand ought to watch its path, but with a simple, tranquil vigilance confined to the present moment, and without restlessness from love of self. Its attention should be continually directed to the will of God, in order to fulfill it every instant, and not be engaged in reflex acts upon itself in order to be assured of its state when God prefers it should be uncertain.

We never watch so diligently over ourselves as when we walk in the presence of God, as he commanded Abraham. And, in fact, what should be the end of all our vigilance? To follow step by step the will of God. He who conforms to that in all things watches over himself and sanctifies himself in everything. If then we never lose sight of the presence of God we should never cease to watch, and always with a simple, lovely, quiet, and disinterested vigilance; while, on the other hand, the watchfulness which is the result of a desire to be assured of our state is harsh, restless, and full of self.

In addition to the presence of God and a state of recollection we may add the examination of conscience according to our need, but conducted in a way that grows more and more simple, easy, and destitute of restless self-contemplations. We examine ourselves not for our own satisfaction, but to conform to the advice we receive, and to accomplish the will of God.

We must silence every creature, including self, that in the deep stillness of the soul we may perceive the ineffable voice of the Bridegroom. We must lend an attentive ear, for his voice is soft and still and is only heard of those who listen for nothing else. How rare is it to find a soul still enough to hear God speak! The least reserve, the slightest self-reflective act, the most imperceptible fear of hearing too clearly what God demands, interferes with the interior voice.

INDEPENDENCE.

Do not suffer yourself to get excited by what is said about you. Let the world talk. Do you strive to do the will of God; as for that of men, you could never succeed in doing it to their satisfaction, and it is not worth the pains.

Let the water flow beneath the bridge. Let men be men, that is to say, weak, vain, inconsistent, unjust, false, and presumptuous; let the world be the world still; you cannot prevent it. Let everyone follow his own inclination and habits; you cannot recast them, and the best course is to let them be as they are and bear with them. Do not think it strange when you witness unreasonableness and injustice; rest in peace in the bosom of God; he sees it all more clearly than you do, and yet permits it. Be content to do quietly and gently what it becomes you to do, and let everything else be to you as though it were not.

As long as the world is anything to us, so long our freedom is but a word, and we are as easily captured as a bird whose leg is fastened by a thread. He seems to be free; the string is not visible, but he can only fly its length, and he is a prisoner.

Do not be vexed at what people say. Let them speak, while you endeavor to do the will of God. A little silence, peace, and communion with God will compensate you for all the injustice of men. We must love our fellow-beings without depending upon their friendship. They leave us, they return, and they go from us again. Let them go or come; it is the feather blown about by the wind. Fix your attention upon God alone in your connection with them. It is he alone who, through them, consoles or afflicts you.

Possess your soul in patience. Renew often within you the feeling of the presence of God, that you may learn moderation. There is nothing truly great but lowliness, charity, fear of ourselves, and detachment from the dominion of sense. Accustom yourself gradually to carry prayer into your daily occupations. Speak, move, act in peace, as if you were in prayer. Do everything without eagerness, as if by the Spirit of God. As soon as you perceive your natural impetuosity impelling you retire into the sanctuary where dwells the Father of Spirits; listen to what you there hear; and then neither say nor do anything but what he dictates in your heart. You will find that you will become more tranquil; that your words will be fewer and more to the purpose, and that with less effort you will accomplish more good. When the heart is fixed on God it can easily accustom itself to suspend the natural movements of ardent feeling, and to wait for the favorable moment when the voice within may speak. This is the continual sacrifice of self, and the life of faith. This death of self is a blessed life; for the grace that brings peace succeeds to the passions that produce trouble. Endeavor to acquire a habit of looking to this light within you; then all your life will gradually become a prayer. You may suffer, but you will find peace in suffering.

THE FAULTS OF OTHERS.

Perfection is easily tolerant of the imperfections of others; it becomes all things to all men. We must not be surprised at the greatest defects in good souls, and must quietly let them alone until God gives the signal of gradual removal; otherwise we shall pull up the wheat with the tares.

They who correct others ought to watch the moment when God touches their hearts; we must bear a fault with patience till we perceive his Spirit reproaching them within. We must imitate him who gently reproves, so that they feel that it is less God that condemns them than their own hearts. When we blame with impatience because we are displeased with the fault it is a human censure, and not the disapprobation of God. It is a sensitive self-love that cannot forgive the self-love of others. The more self-love we have the more severe our censures. There is nothing so vexatious as the collisions between one excessive self-love and another still more violent and sensitive. The passions of others are infinitely ridiculous to those who are under the dominion of their own. The ways of God are very different. He is ever full of kindness for us, he gives us strength, he regards us with pity and condescension, he remembers our weakness, he waits for us. The less we have ourselves the more considerate we are of others.

I am very sorry for the imperfections you find in human beings, but we must learn to expect but little from them; this is the only security against disappointment. We must receive from them what they are able to give us, as from trees the fruits that they yield. God bears with imperfect beings even when they resist his goodness. We ought to imitate this merciful patience and endurance. It is only imperfection that complains of what is imperfect. The more perfect we are the more gentle and quiet we become toward the defects of others.

The defects of our neighbor interfere with our own; our vanity is wounded by that of another; our own haughtiness finds our neighbor's ridiculous and insupportable; our restlessness is rebuked by the sluggishness and indolence of this person; our gloom is disturbed by the gayety and frivolity of that person, and our heedlessness by the shrewdness and address of another. If we were faultless we should not be so much annoyed by the defects of those with whom we associate. If we were to acknowledge honestly that we have not virtue enough to bear patiently with our neighbor's weaknesses we should show our own imperfection, and this alarms our vanity. We therefore make our weakness pass for strength, elevate it to a virtue, and call it zeal. For is it not surprising to see how tranquil we are about the errors of others when they do not trouble us, and how soon this wonderful zeal kindles against those who excite our jealousy or weary our patience?

HUMILITY.

The foundation of peace with all men is humility. Pride is incompatible with pride; hence arise divisions in the world. We must stifle all rising jealousies, all little contrivances to promote our own glory, vain desires to please, or to succeed, or to be praised, the fear of seeing others preferred to ourselves, the anxiety to have our plans carried into effect, the natural love of dominion, and desire to influence others. These rules are soon given, but it is not so easy to observe them. With some people not only pride and hauteur render these duties very difficult, but great natural sensitiveness makes the practice of them nearly impossible, and, instead of respecting their neighbor with a true feeling of humility, all their charity amounts only to a sort of compassionate toleration that nearly resembles contempt.

Humility is the source of all true greatness; pride is ever impatient, ready to be offended. He who thinks nothing is due to him never thinks himself ill-treated; true meekness is not mere temperament, for this is only softness or weakness.

There is no true and constant gentleness without humility; while we are so fond of ourselves we are easily offended with others. Let us be persuaded that nothing is due to us, and then nothing will disturb us. Let us often think of our own infirmities and we shall become indulgent toward those of others.

MODERATION.

The best and highest use of your mind is to learn to distrust yourself; to renounce your own will and to submit to the will of God; to become as a little child. It is not of doing different things that I speak, but of performing the most common actions with your heart fixed on God, and as one who is accomplishing the end of his being. You will act as others do, except that you will never sin. You will be a faithful friend, polite, attentive, complaisant, and cheerful, at those times when it is becoming in a true Christian to be so. You will be moderate at table, moderate in speaking, moderate in expense, moderate in judging, moderate in your diversions; temperate even in your wisdom and foresight. It is this universal sobriety in the use of the best things that is taught us by the true love of God. We are neither austere, nor fretful, nor scrupulous, but have within ourselves a principle of love that enlarges the heart and sheds a gentle influence upon everything; that, without constraint or effort, inspires a delicate apprehension lest we should displease God, and that arrests us if we are tempted to do wrong.

VARIOUS ADVICES.

Peace in this life springs from acquiescence even in disagreeable things, not in an exemption from suffering.

Let us do good according to the means which God has given us, with discretion, with courage, and with perseverance. We shall find occasions to do good everywhere; they surround us; it is the will that is needed. The deepest solitudes, when we seem to have the least communication with others, will furnish us with means of doing good to our fellow-beings, and of glorifying him who is their Master and ours.

A life of faith produces two things: First, it enables us to see God in everything; secondly, it holds the mind in a state of readiness for whatever may be his will. This continual, unceasing dependence on God, this state of entire peace and acquiescence of the soul in whatever may happen, is the true, silent martyrdom of self.

We cannot always be doing a great work, but we can always be doing something that belongs to our condition. To be silent, to suffer, to pray when we cannot act, is acceptable to God. A disappointment, a contradiction, a harsh word received and endured as in his presence, is worth more than a long prayer; and we do not lose time if we bear its loss with gentleness and patience, provided the loss was inevitable and was not caused by our own fault.

The best proof that we are influenced by the Spirit of God is, first, when the action itself is pure and conformable to the perfection of his laws; secondly, when we perform it simply, tranquilly, without eagerness to do it, contented if it is necessary to relinquish it; thirdly, when, after the work is done, we do not seek by unquiet reflections to justify the action even to ourselves, but are willing it should be condemned, or to condemn it ourselves, if any superior light discovers it to be wrong; and when, in fine, we do not appropriate the action to ourselves, but refer it to the will of God; fourthly, when this work leaves the soul in its simplicity, in its peace, in its own uprightness, in humility, and in self-forgetfulness.

The soul in the state of pure love acts in simplicity. Its inward rule of action is found in the decisions of a sanctified judgment. These decisions, guided as they are by a higher power, based upon judgments that are free from self-interest, are the voice of God in the soul. They may not always be *absolutely* right, because our views and judgments, being limited, can extend only to things in part; but they may be said to be *relatively* right; they conform to things so far as we are permitted to see them and understand them, and convey to the soul a moral assurance that, when we act in accordance with them, we are doing as God would have us do. But we must be sure that the soul is free from any selfish bias whatever.

As things are in the present life, those who are wholly devoted to God may suffer in the inferior part (the natural appetites, propensities, and affections), and may be at rest in the superior (the judgment, the moral sense, and the will). Their wills may be in harmony with the divine will; they may be approved in their judgments and conscience, and at the same time may suffer greatly in their physical relations and in their natural sensibilities. In this manner Christ, upon the cross, while his will remained firm in its union with the will of his heavenly Father, suffered much through his physical system; he felt the painful longings of thirst, the pressure of the thorns, and the agony of the spear. He was deeply afflicted, also, for the friends he left behind him and for a dying world. But in his inner and higher nature, where he felt himself sustained by the secret voice uttered in his sanctified conscience and in his unchangeable faith, he was peaceful and happy.

Evil is changed into good when it is received in patience through the love of God; while good is changed into evil when we become attached to it through the love of self.

With the exception of sin, nothing happens in this world out of the will of God. It is he who is the author, ruler, and bestower of all; he has numbered the hairs of our head, the leaves of every tree, the sand upon the seashore, and the drops of the ocean.

This is the whole of religion: to get out of self in order to get into God.

One of the cardinal rules of the spiritual life is that we are to live exclusively in the present moment, without casting a look beyond.

We must imitate Jesus—live as he lived, think as he thought, and be conformed to his image, which is the seal of our sanctification. To be a Christian is to be an imitator of Jesus Christ. In what can we imitate him if not in his humiliation? Nothing else can bring us near to him. We may adore him as omnipotent, fear him as just, love him with all our heart as good and merciful, but we can only imitate him as humble, submissive, poor, and despised.

What men stand most in need of is the knowledge of God. It is not astonishing that men do so little for God, and that the little which they do costs them so much. They do not know him; scarcely do they believe that he exists. If he were known he would be loved.

Thou causest me clearly to understand that Thou makest use of the evils and imperfections of the creature to do the good which thou hast determined beforehand. Thou concealest thyself under the importunate visitor who intrudes upon the occupation of thine impatient child, that he may learn not to be impatient, and that he may die to the gratification of being free to study or work as he pleases. Thou availest thyself of slanderous tongues to destroy the reputation of thine innocent children, that, beside their innocence, they may offer thee the sacrifice of their too highly cherished reputation. By the cunning artifices of the envious, thou layest low the fortunes of those whose hearts were too much set upon their prosperity. Thus thou mercifully strewest bitterness over everything that is not thyself, to the end that our hearts, formed to love thee and to exist upon thy love, may be, as it were, constrained to return to thee by a want of satisfaction in everything else.

THOMAS C. UPHAM.

The Rev. Thomas Cogswell Upham, D.D., was born in Deerfield, N. H., January 30, 1799, and died in New York, April 2, 1872. Graduating at Dartmouth College in 1818, and Andover Theological Seminary in 1821, he was for a time an assistant instructor in the latter school, and for two years was pastor of the Congregational church in Rochester, N. H. But his life was mainly spent at Bowdoin College, Maine, where he was professor of mental and moral philosophy from 1825 to 1867.

He was a prolific writer. His *Elements of Mental Philosophy*, in two volumes, 1839, was for a long time a standard work. It is with his religious productions, however, that we have chiefly to do, and it is by these, we think, that he will be longest known.

The full title of the first (issued in 1843, "to aid in promoting holy living") was, *Principles of the Interior or Hidden Life, designed particularly for the Consideration of Those who are Seeking Assurance of Faith and Perfect Love.* In 1845 appeared the second, entitled *The Life of Faith, in three parts, embracing some of the Scriptural Principles or Doctrines of Faith, the Power or Effects of Faith in the Regulation of Man's Inward Nature, and the Relation of Faith to the Divine Guidance.* In this same year, 1845, was issued the *Life of Madame*

Catharine Adorna, including some Leading Facts and Traits in her Religious Experience, together with Explanations and Remarks tending to Illustrate the Doctrine of Holiness. Next, in two volumes, 1846, came the *Life and Religious Opinions and Experience of Madame de la Mothe Guyon; together with some Account of the Personal History and Religious Opinions of Fénelon, Archbishop of Cambray.* Along the same general line, in 1851, came *A Treatise on Divine Union, designed to point out some of the Intimate Relations between God and Man in the Higher Forms of Religious Experience.* This passed through five editions in a few years. Of less importance are *Religious Maxims* (1854), *Method of Prayer* (1859), *Christ in the Soul* (1872), and *The Absolute Religion* (1872).

The characteristics of all these books are much the same. The author displays in them all the power of close analysis and clear statement that might be expected from a professor of mental philosophy. He shows also an intimate acquaintance with the great devotional writers of the past, quotations from whom have been given in these pages. Very great numbers of people have been exceedingly benefited by reading these works. They belong to a past generation, and are now for the most part out of print, but occasionally a copy can be found. The extracts we furnish will suffice to indicate the style. And though the ideas are not specifically different from those already presented, a somewhat different putting will lend them freshness, and repetition will emphasize the truth.

EVERY EVENT A PROVIDENCE.

Whatever takes place, sin only excepted, is to be regarded as expressive, in some important and positive sense, of the will of the Lord. The controlling presence of the Almighty is there. God is in it. Whatever takes place, with the exception of sin, is not only a portion in the great series of events, but takes place in accordance with the well-considered and divinely ordered arrangement or plan of things. Accordingly, everything which takes place indicates, all things considered, the mind of God in that particular thing. And hence we may be said to reach, through the divine providences, a portion of the divine mind, and to become acquainted with it.

Until the divine intimations within are cleared up and illustrated by the subsequent openings of providence, it seems to me to be the duty of Christians to remain in the attitude of patient expectation and of humble and quiet faith. This doctrine strikes at the root of too great eagerness of spirit, and of all inordinate self-activity. He who would walk with God must walk in God's order. God not only requires us to obey and serve him, but to obey and serve him in his own time and way. A soul wholly devoted to God will always endeavor to move calmly, yet firmly and exactly, in the blessed order of the divine providences; neither prematurely and excitedly hastening in advance, nor yet sluggishly and carelessly lagging behind.

The existence of an undue eagerness and excitement of spirit is an evidence that we are, in some degree, afraid to trust God, and that we are still too much under the influence of the life of nature. So that to cease from the activity of nature, when properly understood, seems to be nothing more nor less than to cease from the spirit of self-wisdom, self-seeking, and self-guidance, and thus to remain in submissive and peaceful simplicity and disengagement of spirit, in order that God may enter in, and may guide us by the wisdom of his own divine inspiration.

It is the rejection of the doctrine of providence considered as entering into particulars which constitutes one of the great evils, the practical atheism, perhaps we may call it, of the age in which we live. It is true, undoubtedly, that men, with but few exceptions, admit the existence of a God; but they do not admit, except in a very mitigated and imperfect sense, his presence and supervision.

As the law of providence is only another expression for God's will as that will is exhibited in connection with his providences, the man who lives in conformity with providence necessarily lives in conformity with God. It is only when we are in this position that we may be said to walk with God; and walking with God is union with God. To be in harmony with God's providence we must be in harmony with everything, not excepting the material world. It is true that things inanimate have no life in themselves, but they are the residence of a living mind. We might almost say, in a mitigated sense of the terms, that everything, not excluding objects the most remote from moral intelligence, becomes God to us. There is no grass, no flower, no tree, no insect, no creeping thing, no singing bird, nothing which does not bring God with it, and in such a manner that the thing which we behold becomes a clear and bright revelation of that which is invisible.

The event, painful as it is and criminal as it is under some circumstances, is nevertheless a manifestation of God; and not of a God absent, but of a God present. And happy is the man that can receive this. To be out of harmony with these things, acts, and events which God in his providence has seen fit to array around us—that is to say, not to meet them in a humble, believing, and thankful spirit—is to turn from God.

Everything which occurs, with the exception of sin, takes place—and yet without infringing on moral liberty—in the divinely appointed order and arrangement of things, and is an expression, within its own appropriate limits, of the divine will. And consequently, in its relations to ourselves personally and individually, it is precisely that condition of things which is best suited to try and to benefit our own state. On a moment's reflection it will be seen that this important principle raises us at once above all subordinate creatures, and places us in the most intimate connection with God himself. It makes the occurrences of every moment, to an important extent, a manifestation of God's will, and consequently, in every such occurrence, it makes God himself essentially present to us. Every event coming within the range of our cognizance necessarily brings God and our souls together. And it naturally follows from this view that everything which takes place, whatever it may be, inasmuch as it is a revelation, within its appropriate limits, of God's presence and God's will, should be met in the spirit of acquiescence, meekness, and entire resignation.

Faith aids the soul by calling to its remembrance, and by establishing its belief, that all events, including what are called evils, make a part in God's providences. We sometimes err by limiting the sphere of providential arrangements. These arrangements extend to everything which does not interfere with the claims of moral agency. They include mind as well as matter. It is an important truth, though not always recognized, that mental trials, as well as those which are purely physical, may have their origin from God.

The form of faith which is especially necessary in order to live the life of faith is that which makes God present, moment by moment, either permissively or causatively, in any and all events which take place. O that we might learn the great lesson (the lesson absolutely indispensable to him who would experience the highest results of the inward life) of beholding God, either in his direct efficiency or his permissive and controlling guardianship, as present in all things, whether high or low, of whatever name or nature! Without taking this view of his presence we deprive ourselves of that great Center where the soul finds rest. We are tossed and agitated by passing events. Everything is perplexed, mysterious, and hopeless.

BEST PROOF OF PERFECT LOVE.

When there is an entire and cordial acquiescence in the will of God, both to do and to suffer, we have the most important and satisfactory mark that our love is perfect. The nature of the human mind is such that we never can have an entire and cordial acquiescence in the will of God in all things without an antecedent approval of and complacency in his character and administration.

It was one of the sayings of the devout Francis Xavier that "the perfection of the creature consists in willing nothing but the will of the Creator." What other idea of perfection of love can we have?

This is the true mark of perfection in Christian love, namely, an entire coincidence of our own wills with the will of God; a full and hearty substitution of the divine mind in the place of our own minds; the rejection of the natural principle of life, which is love terminating in self, and the adoption of the heavenly principle of life, which is love terminating and fulfilled in God; in other words, the expulsion of self from the heart, and the enthronement of God there as its everlasting sovereign. This view, so important practically as well as theologically, seems to be confirmed by what the Saviour says of himself in a number of passages (John vi, 38; John iv, 34; Heb. x, 9; Mark iii, 34, 35; Matt. vii, 21).

THE IMAGE OF CHRIST.

Some of the traits of character which are conspicuous in the life of our Saviour: He was a man of sympathy. He was susceptible of, and actually formed, to some extent, personal friendships and intimacies. He exhibited and valued intellectual culture. I have sometimes thought that persons of flighty conceptions and vigorous enthusiasm would regard the Saviour, if he were now on the earth, as too calm and gentle, too thoughtful and intellectual, too free from impulsive and excited agitations, to be reckoned with those who are often considered the most advanced in religion.

The life of the Saviour was characterized by the

spirit of entire consecration. He lived by simple faith. He never doubted. Faith sustained him in trial as well as in duty; in the depths of affliction as well as in the active labors of his ministry. He was a man of prayer. He was conscientiously and strictly faithful in whatever the Father committed into his hands to do. "He pleased not himself." In the various companies in which he mingled he never forgot the great mission on which he came. He was not, however, prematurely zealous and obtrusive. He realized that everything, when done in accordance with the will of his heavenly Father (a will which can never be at variance with the highest rationality), must necessarily have its right time and place.

He exhibited in his daily deportment a very meek, humble, and quiet disposition of mind. In the possession of the inestimable trait of meekness and quietness of spirit let all who seek the highest degree of purification and sanctification of heart be imitators of the example of Jesus Christ. The life of the Saviour was characterized by a proportionate fitness or symmetry in all its parts.

In all cases of true holiness, without exception, there must be, and there is, the image of Christ at the bottom. The soul becomes an "infant Jesus," and like its all-perfect prototype it will grow in "wisdom and in stature, and in favor with God and man."

SPIRITUAL FREEDOM.

The person is not in the enjoyment of true liberty of spirit who is wanting in the disposition of accommodation to others in things which are not of special importance. And this is the case when we needlessly insist upon having everything done in our own time and manner; when we are troubled about little things which are in themselves indifferent, and think, perhaps, more of the position of a chair than of the salvation of a soul; when we find a difficulty in making allowance for the constitutional differences in others which it may not be either easy or important for them to correct; when we find ourselves disgusted because another does not express himself in entire accordance with our principles of taste; or when we are displeased and dissatisfied with his religious or other performances, although we know he does the best he can. All these things, and many others like them, give evidence of a mind that has not entered into the broad and untrammeled domain of spiritual freedom.

The person who is disturbed and impatient when events fall out differently from what he expected and anticipated is not in the enjoyment of true spiritual freedom. In accordance with the great idea of God's perfect sovereignty the man of a religiously free spirit regards all events which take place, sin only excepted, as an expression, under the existing circumstances, of the will of God. And such is his unity with the divine will that there is an immediate acquiescence in the event, whatever may be its nature, and however afflicting in its personal bearings. His mind has acquired, as it were a divine flexibility, in virtue of which it accommodates itself with surprising ease and readiness to all the developments of Providence, whether prosperous or adverse.

The person who enjoys true liberty of spirit is the most deliberate and cautious in doing what he is most desirous to do. This arises from the fact that he is very much afraid of being out of the line of God's will and order. He distrusts and examines closely all strong desires and strong feelings generally, especially if they agitate his mind and render it somewhat uncontrollable. Not merely because the feelings are strong, but because there is reason to fear that some of nature's fire has mingled with the holy and peaceable flame of divine love.

Freedom consists not in having things in our own way, but in the right way, which is God's way. And this includes not only the thing done, but the manner of doing it, and also the time. True liberty of spirit is found only in those who, in the language of De Sales, "keep the heart totally disengaged from every created thing, in order that they may follow the known will of God."

Spiritual liberty consists in passively, yet intelligently and approvingly, following the leadings of the Holy Ghost. It is like a little child that reposes in simplicity and in perfect confidence on the bosom of its beloved mother. It implies, with the fact of entire submission to God, the great and precious reality of interior emancipation. He who is spiritually free is free in God. And he may, perhaps, be said to be free in the same sense in which God is, who is free to do everything right and nothing wrong.

ABSOLUTE SURRENDER.

The prostration of our own will, in such a sense that it shall not in any respect oppose itself to the will of God, seems to be the completion or consummation of those various interior processes by which the heart is purified. The moment our faith in God wavers, that moment we begin to form our own plans and set up our own wills. So that we can have no hesitancy in saying that a will perfectly coincident with the will of God is at the same time the natural result and the highest evidence of a sanctified heart. When the will in its personal or self-interested operation is entirely prostrated, so that we can say with the Saviour, "Lo, I come to do thy will," then the wall of spiritual separation is taken away, and the soul may be said, through the open entrance, to find a passage, as it were, into God himself, and to become one with him in a mysterious but holy and glorious union.

The person whose will is entirely subdued, so as

to be one with the divine will, will discover an unruffled meekness and quietness of spirit when called in the divine providence to endure the smaller and more frequent inconveniences and vexations of life. Nor is the evidence which is thus presented of an entire subjection of the will to be regarded as inconsiderable and unimportant. It is truly sad and humiliating to see many who, in the comparative sense of the term, are good Christians, that are, nevertheless, uneasy, and are inwardly and outwardly vexed, on many trivial occasions.

The man whose will has passed from his own unsafe keeping into the high custody of a divine direction has no disposition to complain when God, in his holy providence, in depriving him of health, of property, and friends, has laid waste his fairest earthly prospects. He endures also in quietness and silence of spirit misrepresentations and persecutions. Strong in a faith which has become habitual to him, he sees everything in its relation to the divine mind. He regards the persecutions he endures as the lot which God has appointed to him, and as such he rejoices in it.

The man who has experienced the practical annihilation of his own will does everything and suffers everything precisely in the order of God's providence. It is the present moment, considered as indicating the divine arrangement of things, which furnishes the truest and safest test of character. It is necessary to keep our eye fixed upon God's order. We must do this in relation to our place and situation in life, whatever it may be; not murmuring at our supposed ill lot, not giving way to any eager desires of change, but remaining quietly and humbly just where God has seen fit to place us.

DEGREES OF DIVINE UNION.

The first degree may be described as union with the divine will *in submission.* It is the union of simple acquiescence rather than of positive desire; the union of submission to suffering rather than of love to suffering. The fact of obedience, however sincere and true the obedience itself may be, does not prevent their saying, with equal truth, that it is hard for nature to yield to it. There is submission in fact, but a submission which costs a struggle in the beginning, and watchfulness and struggles in the maintenance of it.

The second degree may be described as union with the divine will *with choice.* We not only submit, but submission is our pleasure, our delight. The endurance of loss and suffering is not, and cannot ordinarily be, so great as to prevent a true and substantial joy of the heart. It is said of the early Christians not merely that they submitted to suffering with patience, but that they rejoiced that they were accounted worthy to suffer for the name of Jesus (Acts v, 41).

This last state of mind may assume a new character, and may present the union of the will in a new aspect, by becoming invigorated and perfected *by habit.* It may ultimately become so well established and strong that the effect of antecedent evil habits, which generally remains for a long time and greatly perplexes the full sway of holiness in the heart, shall be done away entirely. And this is not all. In the course of time our perceptions of the transcendent beauty and excellence of the will of God may become so increased in clearness and strength that the pleasure of doing and suffering his will, increased in the same proportion, may entirely absorb and take away our sense of suffering. The suffering will be lost in the joy. "Death," a name which includes all temporal evil, "will be swallowed up in victory."

RECEIVING BY FAITH.

On the true doctrine of holy living, namely, by faith, we go to God in the exercise of faith, believing that he will hear; and we return from him in the exercise of the same faith, believing that he *has* heard, and that the answer exists and is registered in the divine mind, although we do not know what it is, and perhaps shall never be permitted to know.

If we truly and humbly ask for wisdom to guide us, and at the same time, of course, employ all those rational powers which God has given us, it becomes our privilege and our duty, in accordance with the doctrines of the life of faith, to believe fully and firmly that God does in fact answer, and that in the sanctified exercise of the powers which are given us we truly have that degree of wisdom which is best for us in the present case. This, whether we are conscious of any new light on the subject or not. Even if we are left in almost total ignorance on the topic of our inquiry, we have the high satisfaction of knowing that we are placed in this position because God sees that a less degree of light is better in our case than a greater.

The system which requires a present and visible or ascertained answer, in distinction from the system of faith, which believes that it has an answer but does not require God to make it known till he sees best to make it known, is full of danger. It tends to self-confidence, because it implies that we can command God, and make him unlock the secrets of his hidden counsels whenever we please. It tends to self-delusion, because we are always liable to mistake the workings of our own imaginations, or our own feelings, or the intimations of Satan, for the true voice of God. It tends to cause jealousies and divisions in the Church of Christ, because he who supposes that he has a specific or known answer, which is the same, so far as it goes, as a specific revelation, is naturally bound and led by such supposition, and thus is oftentimes led to strike out a course for himself which is at variance with the feelings and judgments of his brethren. Incalculable are the evils which, in every age of the Christian history, have resulted from this source.

On the contrary, the disposition to know only what God would have us know, and to leave the dearest objects of our hearts in the sublime keeping of the general and unspecific belief that God is now answering our prayers in his own time and way, and in the best manner, involves a present process of inward crucifixion which is obviously unfavorable to the growth, and even existence, of the life of self.

Faith in its relation to the subject of it is truly a light in the soul, but it is a light that shines only upon *duties,* and not upon results or events. It tells us what is now to be done, but it does not tell us what is to follow. And accordingly it guides us but a step at a time. And when we take that step under the guidance of faith we advance directly into a land of surrounding shadows and darkness. Like the patriarch Abraham, we go, not knowing whither we go, but only that God is with us. In man's darkness we nevertheless walk and live in God's light, a way of living which may well be styled blessed and glorious, however mysterious it may be to human vision. Indeed, it is the only life worth possessing, the only true life. "Believe in the Lord your God, so shall ye be established."

LIVING BY THE MOMENT.

We are not at liberty to attach ourselves strongly to plans of action. We ought to sit loosely to everything except the present moment. We ought not to permit our affections to become enlisted, as they are very apt to be. We should enter upon the plan in accordance with God's will; we should advance step by step in accordance with his will; and without the least emotion of disappointment or displeasure we should stop in accordance with his will; which we cannot well do if we let our affections go in advance of the divine moment, which is the present moment, and cleave to objects which have not as yet received the divine sanction.

No man lives well who lives out of the will of God. No man lives in the will of God who anticipates the divine moment, or moment of actual duty, by making up a positive decision before it arrives or by delaying a decision until after its departure. If, therefore, we would live in the will of God we must conform to that beautiful and sacred order in which his will is made known; we must live by the moment.

This doctrine keeps the mind fixed to God alone. Every moment presents our blessed Maker before us, with the facts of his providence all arranged and convergent to one point, and requiring of us as moral agents a prompt decision. God is in that moment as it arrives; his unseen presence is embodied in that small point of time; he speaks to us in the still small voice; if we hear, and reply with correspondent heart and action, it is well; if we do not listen and obey he is gone from us; and an eternity to come cannot remedy the loss of that one moment.

It is a result of these principles that they preserve us from the very considerable evil of reflex acts of mind; that is to say, of frequent and unnecessary returns of the mind upon itself in the form of self-inquiry, of self-condemnation, or of self-gratulation, and in other ways which might be mentioned. This result seems to follow from the fact that, on the system of living by the moment, the mind always has before itself a present object, and that the object fully occupies and absorbs the mind, because God himself is present in it.

GOD'S GUIDANCE.

In many cases, where the motives which are presented are various and the paths of action are divergent, it is not easy for us to know, with absolute certainty, what course of action will most fully accord with the divine will. Constituted as we are at present, we may well pronounce it impossible to have such knowledge except by means of a specific revelation given in each case. And we may even go further and say, it is not the design of our heavenly Father that in matters of this kind we should always have a knowledge which is positive, and should always walk in a vision which is open. This is not God's plan of action. We must, in a considerable degree at least, live by faith.

The prayer for divine direction, offered up in the spirit of consecration, which implies a heart wholly given to God, and offered up also in entire faith, which receives the promises of God without wavering, necessarily involves the result that the course taken, whether it be conformed to natural wisdom or not, and is attended with the best natural results or not, is morally the right course, and is entirely acceptable to God. A man in that state of mind may commit a physical or prudential error; he may perhaps take a course which will be followed by the loss of his property, or an injury to his person, but he cannot commit a moral error. That is to say, he cannot commit an error which, under the adjustments and pledges of the Gospel, will bring him into a state of moral condemnation and separate him from God's favor.

In acting in accordance with the results which we thus obtain we always and necessarily accomplish the will of God. We know his will, while in a certain sense we may be said to be ignorant of it; because it is his will that we should live and act by faith without knowledge. "I adore all thy purposes," says Fénelon, "without knowing them." This is the great work of holiness, to do the will of

God, while we know it, and can know it, only in part. Living by faith without knowledge is living in the truest divine light. When we are led in the way of faith we are led by God himself; and it is impossible for God, by means of spiritual operations, to lead his people in a way which is contrary to his will.

To the question, How shall we know the will of God specifically, or in particular cases? our answer is that God always meets us with a specific revelation of his will in the events or providences of the present moment. In other words, the events of God's providence, just so far as they give us information at all, are to be regarded as an expression of his will. And so far as they do not give us information of themselves they furnish a basis of information which may be deduced from them.

Consequently we are not at liberty to pronounce what the will of God is, in relation to a course of action, until the present moment, as we may conveniently designate the precise period of action, has come. In order to know what is right and duty we must have *all* the facts; but no moment, antecedent to the present moment, or the precise moment of action, can give them. This is a state of things which has the obvious advantage of being opposed to self-confidence and rash judgments, and of being favorable to forbearance, charity, and humility. Hence it is that very holy men, in a multitude of cases, defer their judgments, while others, less holy, are prompt in deciding.

RELIGIOUS MAXIMS.

In whatever you are called upon to do endeavor to maintain a calm, collected, and prayerful state of mind.

Let the heart be fully united with the will of God, and we shall be entirely contented with those circumstances in which Providence has seen fit to place us, however unpropitious they may be in a worldly point of view. He who gains the victory over himself gains the victory over all his enemies.

It may sometimes be practically important to make a distinction between a renunciation of the world and a renunciation of ourselves. A mere crucifixion of the outward world may still leave a vitality and luxuriance of the selfish principle; but a crucifixion of self necessarily involves the crucifixion of everything else.

It is one among the pious and valuable maxims which are ascribed to Francis de Sales, "A judicious silence is always better than truth spoken without charity." The very undertaking to instruct or censure others implies an assumption of moral or intellectual superiority. It cannot be expected, therefore, that the attempt will be well received unless it is tempered with a heavenly spirit.

Perhaps we may say it is the highest attainment

of the soul (certainly it is the foundation of the highest or perfect state), that of entire and unwavering confidence in God.

Always make it a rule to do everything in the best manner, and to the best of your ability. An imperfect execution of a thing which we might have done better is not only unprofitable, but it is a *vicious* execution; it is morally wrong.

A fixed, inflexible will is a great assistance in a holy life. He who is easily shaken will find the way of holiness difficult, perhaps impracticable. Ye who walk in the narrow way, let your resolution be unalterable.

When on a certain occasion the pious Fénelon, after having experienced much trouble and persecution from his opposers, was advised by some one to take greater precautions against the artifices and evil designs of men, he made answer, in the true spirit of a Christian, "Let us die in our simplicity." He that is wholly in Christ has a oneness and purity of purpose altogether inconsistent with those tricks and subterfuges which are so common among men. He walks in broad day. He goes forth in the light of conscious honesty. He is willing that men and angels should read the very bottom of his heart. He has but one rule: "My Father, what wilt thou have me to do?"

It is important to make a distinction between sorrow and impatience. We may feel sorrow without sin, but we can never feel impatience without sin. Impatience always involves a want of submission; and he who is wanting in submission, even in the smallest degree, is not perfect before God.

Many profess religion; many, we may charitably hope, possess religion; but few, very few, if we may judge from appearances, are aiming with all their powers at perfection in religion. Nevertheless it is only upon this last class that the Saviour looks with unmingled approbation.

If we would walk perfectly before God we must endeavor to do common things, such as are of every day's occurrence and of but small account in the eyes of the world, in a perfect manner.

It will help us to ascertain whether we are truly humble if we inquire whether we are free from the opposites of humility. The opposites of a humble state of mind are impatience, uneasiness, a feeling that something—perhaps much—depends on ourselves, undue sensitiveness to the praise and reproof of men, and censoriousness.

A state of suffering furnishes the test of love. When God is pleased to bestow his favors upon us, how can we tell whether we love him for what he *is,* or for what he *gives?* But when in seasons of deep and varied afflictions our heart still clings to him as our only hope and joy, we may well say, "Thou knowest all things; thou knowest that I love thee."

A consecration deliberately made, including all

our acts, powers, and possessions, of body, mind, and estate; made without any reserve either in objects, time, or place; embracing trial and suffering as well as action; never to be modified and never to be withdrawn, and which contemplates its fulfillment in divine and not in human strength, necessarily brings one into a new relationship with God, of the most intimate, interesting, and effective nature.

We are not to desire anything whatever out of the will of God. In other words, if we find a preference or choice in ourselves, in such a manner as to lead us to desire one thing rather than another irrespective of the will of God, we may justly conclude that the state of mind of which we are then the subjects is a selfish and natural state, and not a truly religious and divine state. It is to be rejected; and the mind is to remain without desire until the will of God can be revealed and take effect in us.

Quietness of spirit, originating in the operations of divine grace, is the sign of truth or rectitude of spirit, and also of a right cause of action. And, on the other hand, a spirit disturbed, a spirit in a state of agitation, is the sign of a wrong done or proposed to be done. Accordingly, in any proposed course of action, if it cannot be entered upon with entire quietness of spirit, with a soul so entirely calm that, in its measure, it may be said to reflect unbrokenly the image of God, then the probability is that the course proposed to be taken is wrong, or, at least, of a doubtful character; and our true and safe course is to delay until we can obtain further light in regard to it.

He whose life is hid with Christ in God may suffer injustice from the conduct or words of another, but he can never suffer loss. He sees the hand of God in everything.

It is a sign that our wills are not wholly lost in the will of God when we are much in the habit of using words which imply election or choice, such as, I want this, or, I want that; I hope it will be so, or, I hope it will be otherwise.

A holy person often does the same things which are done by an unholy person, and yet the things done in the two cases are infinitely different in their character. The one performs them in the will of God, the other in the will of the creature.

Two things in particular are to be guarded against in all the variety of their forms, namely, creature love and self-will; in other words, dependence upon self and dependence upon our fellow-men.

No person can be considered as praying in sincerity for a specified object who does not employ all the appropriate natural means which he can to secure the object.

The holy mind chooses to be, and loves to be, where it is, and has no disposition or desire to be anywhere else, till the providence of God clearly indicates that the time has come for a removal.

FREDERICK WILLIAM FABER.

IT is not easy to write briefly either about Faber or his books. He had a most fascinating character and a most interesting history. His birth was in the vicarage of Calverley, Yorkshire, England, June 28, 1814. Educated at Oxford, where he obtained a scholarship and a fellowship, he was ordained deacon in 1837 and priest of the Church of England in 1839. He became rector of Elton in 1843, and did his work there with the utmost diligence, producing a great reformation. But for ten or twelve years, partly through the natural bent of his mind, partly through the influences around him, he had been drawn steadily, irresistibly toward Romanism, and at length, after great mental struggles and the most intense desire to do only what was right, he was received, November 17, 1845, into the Roman Catholic Church. He had to make very great sacrifices to carry out his convictions, but the result was peace, and he never doubted that he had been led of the Lord.

His life as a Roman Catholic priest was an extremely busy and useful one. At Birmingham he organized a community called "Brothers of the Will of God." In 1848 he joined the order of St. Philip under Dr. Newman, and from 1849 till his death, in 1862, he was at the head of the London branch, or oratory, of this order. His labors in every possible direction were incessant and marvelously successful, though often broken in upon by serious illness. It is doubtful if any man ever had more of the true spirit of Jesus or brought his life closer to the divine model. He served his Master from love, with all his heart and might. He continually preached Jesus, and him crucified, in the simplest and most earnest way. He threw every ounce of his strength into his efforts to make men good and to extend the spirit of genuine holiness. His humility was most profound, his tenderness and forbearance extraordinary, his love overflowed all bounds of creed or condition. He was one of the most lovable men that ever lived. The charm of his manner, the kindliness of his heart, the genuineness of his sympathy, the brilliancy of his social powers, the ripeness of his worldly wisdom, and the unearthliness of his aims formed one of the rarest of combinations. His life from earliest childhood seems to have been deeply religious. He ever chose the higher path, putting self aside, and seeking only to glorify God.

His writings are divisible into four classes, namely, the works that he translated and edited, the books that he composed in prose, his hymns and poems, and the religious letters in which he replied to the multitude of applications made to him for spiritual counsel. It is perhaps by the hymns that he will longest live; certainly by them more than by anything else he is known to the Protestant world. He was a genuine poet, and the poet of the higher spiritual life more than any other person of modern times. The surpassing beauty and spiritual depth of many of his hymns are recognized by all who have any power to appreciate these things. The religious experience which is voiced in them shows that none but a Christian of maturest piety could have penned them, and the elegance of the style proves that a master hand has been at work. It may well be said that such great gifts, of piety and poetry alike, were rarely before so harmoniously and completely joined.

His great prose works, from which the following extracts are taken, consist of eight solid, close-printed volumes, which were issued in the short space of eight years—1853 to 1860, inclusive. And all this time he was diligently occupied with an amount of other work quite sufficient for an ordinary man, to say nothing of the frequent illnesses and the constant pain under which he had to bear up as best he could. A severe attack of illness infallibly followed the completion of each of his books. We can only give here the bare titles of the eight. *All for Jesus, or the Easy Ways of Divine Love; Growth in Holiness, or the Progress of the Spiritual Life; The Blessed Sacrament, or the Works and Ways of God; The Creator and the Creature, or the*

Wonders of Divine Love; The Foot of the Cross; Spiritual Conferences; The Precious Blood; Bethlehem. They are all notable for the beauty of their style, their accuracy of theological statement, their intimate knowledge of the human heart, and the intensity of the devotion to God which they everywhere inculcate. They sprang at once into great popularity, and hundreds of thousands of copies have been sold in England, Europe, and America.

A small volume on Faber, containing a full sketch of his life, together with all of his best hymns and extended selections from his prose works, was issued a few years ago by the author of this book, and may be procured of him (for fifty cents) by anyone wishing to pursue this fascinating theme.

THE GLORY OF GOD.

Blessed be God! There are many souls to whom his glory is the passion of their lives. The worth of everything to them is simply its capability of glorifying God, and nothing more. Their choice of means and ends is guided by this same propension. Their happiness is their success in this single matter. To them life is a matter of one fact; and all truths resolve themselves into one, and that is the immense worthiness of God to be loved; and it seems as if a necessity were laid upon them to see that he should be infinitely loved even by finite creatures.

When we study our blessed Lord as he is represented to us in the Gospels, nothing, if we may venture to use such an expression, seems so like a ruling passion in him as his longing for his Father's glory.

While the saints differ in almost everything else, there are three things in which they all agree; and these are: (1) Eagerness for the glory of God; (2) Touchiness about the interests of Jesus; (3) Anxiety for the salvation of souls.

THE KNOWLEDGE OF GOD.

Fénelon observed long ago that the general laws of nature are, after all, not so much manifestations of God's presence and perfection as the screen to hide both one and the other. "Why," he asks, "has God established these general laws?" It is to hide under the veil of the regularity and uniform cause of nature his perpetual operation from the eyes of proud and corrupt men, while, on the other hand, he gives to pure and docile souls something which they may admire in all his works.

Men little know how great and good a work it is which they are doing when they increase by ever so little another's knowledge of the Most High. They have not stopped one sin, but hundreds. They have not been the channels of one grace, but of thousands. The knowledge of God is the establishment of Christ's kingdom in the soul. How many would advance in the spiritual life who now stand still because the divine perfections are not preached to them or do not make part of their spiritual reading!

God must be watched in order to be known; and we must watch him on our knees, and in the lowest place within ourselves to which we can sink. Thus we shall learn much if we do not learn all.

The magnificence of God is the abounding joy of life. It is an immense joy to belong to God. It is an immense joy to have such a God belonging to us. Like the joys of heaven, it is a joy new every morning when we wake, as new as if we had never tasted of it before. Like the joys of earth, it is a joy every evening, resting and pacifying to the soul.

All men remember their past lives by certain dates or epochs. Some men date by sorrows, some by joys, and some by moral changes or intellectual revolutions. But the real dates in a man's life are the days and hours in which it came to him to have some new ideas of God. To the thoughtful and the good all life is a continual growing revelation of God. Time itself discloses him. Old truths grow; obscure truths brighten.

To know God and to understand his ways is the great end of life, and to walk in his presence is all sanctity.

TRUST IN GOD.

It is easier to love God than to trust in him. In human things it is not easy to doubt and yet to love, but in divine things it is not uncommon. The greatest defect in our worship of God is want of confidence in him. What can give us more confidence in God than the study of the precious blood! Who can doubt Jesus when he bleeds?

Confidence in God is the only real worship of God. Our confidence is our religion. It is the sweetness of life. It is worth our while to have lived if it were only to have known the delight of trusting in God. Many aim at perfection, and few attain it. In almost every case the reason of the failure is the want of confidence in God.

Meditation on the attributes of God is one of the chief means of acquiring the grace of confidence. In order to have confidence we must know God, know him in Jesus Christ.

Outward temptations help us. They frighten us away from self-trust. They make us better acquainted with our possibilities of sin. A much-tried man is always a man of unbounded faith, and of a confidence in God which looks, to us of lower faith, superstitious in little things and presumptuous in great ones.

We also acquire confidence in God by exercising confidence. It produces itself, and multiplies itself, while it strengthens itself. Direct prayer for the grace is also an obvious means of its increase.

A special devotion to the providence of God, which seems to have possessed the souls of some of the modern saints as a scarcely conscious protest against a false philosophy, is another means of acquiring confidence in God.

But, above all, the habit of working for God only, of doing our good for him, and caring little about its success, and of doing it secretly—which we instinctively do when we do it only for him—is the royal road to confidence in him.

Happy is he who makes one other man trust God more than he did before. He has done a great and influential work in creation. Happy we, if we know how to trust God as he should be trusted.

EDIFICATION.

We must never do anything in order to edify others, for the express purpose of edifying, which we should not have done except to edify them, and in the doing of which the motive of edification is supreme, if not solitary. Edification must never be our first thought. Look out to God, love his glory, hate yourself, and be simple, and you will shine—fortunately without knowing or thinking of it—with a Christlike splendor wherever you go and whatever you do.

We must not make unseasonable allusions to religion, or irritate by misplaced solemnity. An inward aspiration or momentary elevation of the soul to God will often do more, even for others, than the bearing of an open testimony which principle does not require, and at which offense will almost inevitably be taken. A man is annoyed at sacred things when they are unseasonably forced upon him; and thus even a well-meaning importunity may be a source of sin.

We must bear in mind that there are very few who, by standing or advancement, are in any way called upon to correct their brethren, fewer still who are competent to do it sweetly and wisely, and none whose holiness is not tried to the utmost by its perfect discharge.

We may edify our neighbor in two ways: by the mortification of Jesus and by the sweetness of Jesus. Silence under unjust rebukes, abstinence from rash and peremptory judgments, not standing out in an ill-natured and pedantic way for our rights, obliging others unselfishly and with pains and trouble to ourselves, and not exaggerating in an obstinate and foolish manner points where all men have a right to their liberty—these are the ways in which we should practice the mortification of Jesus in our intercourse with others; and, independent of the edification we shall give thereby, the amount of interior perfection which we shall attain by these practices is beyond all calculation. For there is hardly a corrupt inclination, a secret pride, or a fold of self-love which they will not search and purify.

The more earnestly we are striving to form Jesus in our hearts the more will his sweetness transpire through our features without our knowing it. Kind and gentle words, such as those of our dear Lord, are an apostolate in themselves. Our manner, too, must be full of unction, and be of itself a means to attract men to us, and make them love the spirit which animates us. Coldness, absence of interest, an assumption of superiority for some unexpressed reasons, or even an obviousness of condescension, are not unfrequently to be found in pious persons. Sweetness is practiced when we praise all the good we can detect in others, even when it is mingled with what is not so. A man who praises freely but not extravagantly is always influential in conversation, and can use his influence for the cause of God. A critical spirit, on the contrary, amuses by its smartness or frightens by its malignity; but it neither softens, attracts, persuades, nor rules. The practice of putting favorable interpretations upon dubious actions is another exercise of this Christlike sweetness. You will never practice it without having done some missionary work for the glory of God, although you know it not.

LUKEWARMNESS.

Lukewarmness is often nothing more than a clogging up of the avenues of the soul with sins of omission, so that the cool and salutary inundations of grace are hindered. The symptoms of lukewarmness are seven in number: first, a great facility in omitting our exercises of piety; second, negligence in those we do perform; third, a feeling that we are not altogether right with God, joined with an unwillingness to vigorously face the inquiry as to just what is wrong, and to buckle to the triple task of discovery, punishment, and reformation; fourth, habitually acting without any intention at all, good, bad, or indifferent; fifth, carelessness about forming habits of virtue; sixth, contempt of little things, and of daily opportunities; seventh, thinking rather of the good we have done than of the good we have left undone, resting on the past rather than striving for the future, loving to look at people below us rather than at people above us.

Why does God hate lukewarmness so? (Rev. iii, 15, 16.) Because it is a quiet, intentional appreciation of other things over God. It cheapens God, and parts with him secondhand. It pretends friendship; hence it involves the twofold guilt of treachery and hypocrisy. It thus has a peculiar ability to wound God's glory by the scandal it gives. It has God's honor in its power, and treats it shamefully and cruelly. It profanes grace by the indifference with which it misuses it.

Remedies for lukewarmness: The only sure one is never to be lukewarm. Some others that may be mentioned: To quicken faith by meditation on eternal truths; to have fewer things to do; to persevere in our spiritual exercises in spite of dryness and distractions; to talk less, and to mortify the flesh.

PURITY OF INTENTION.

The only important thing in good works is the amount of love which we put into them. The soul of an action is its motive. The power of an action is neither in its size nor in its duration, but in its intention. An intention is pure in proportion as it is loving. What we want is not many actions, but a great momentum in a few actions.

In good deeds we cannot unite number and momentum. We make our election of momentum. Momentum is purity of intention. Purity of intention is love. The saints were men who did less than other people, but who did what they had to do a thousand times better. They threw immense effort into their least actions. Immense efforts cannot help being limited in number.

Have we ever done any one action which we are quite confident was done solely and purely for the love of God? If we have, it has not been often repeated. We are conscious to ourselves that there is a great admixture of earthly motives in our service of God.

There is not a single thing we do all the day long which may not, and that quite easily, be made to advance the glory of God, the interests of Jesus, and the salvation of souls. If the heavenly motive enters into it, that moment it is all filled with God, and becomes a jewel of almost infinite price, with which the Divine Majesty condescends to be well pleased.

We must do all our actions for God, referring them to him by an act of intention. We must momentarily collect ourselves before acting, and try to touch lightly the beginning, middle, and end of each considerable action, and not throw away, as fish too small for the table, the little actions of the day.

One sign that we are really working for God is, if we could say "Yes," did any one suddenly ask us if what we are doing is for God. Another is, if we are not uneasily anxious about the judgments men will pass upon our actions. A third is, if we are not wholly indifferent, but quite tranquil about success. A fourth is, if we take as much pains in private with what we are doing as in public before witnesses. A fifth is, if we are not jealous either of associating others with our works or of their equal or greater success.

SIGNS OF PROGRESS.

Five signs of progress in the spiritual life: (1) If we are discontented with our present state, whatever it may be, and want to be something better and higher, we have great reason to be thankful to God. For such discontent is one of his best gifts, and a great sign that we are really making progress. But we must remember that our dissatisfaction with ourselves must be of such a nature as to increase our humility, and not such as to cause disquietude of mind or uneasiness in our devotional exercises. (2)

It is a sign of growth if we are always making new beginnings and fresh starts. These consist chiefly in two things: first, a renewal of our intention for the glory of God; and, secondly, a revival of our fervor. (3) It is a sign of progress when we have some definite thing in view; for instance, if we are trying to acquire the habit of some particular virtue, or to conquer some besetting infirmity. (4) It is a still greater sign that we are making progress if we have a strong feeling in our minds that God wants something particular from us. (5) I will venture to add that an increased general desire of being more perfect is not altogether without its value as a sign of progress—and that in spite of what I have said of the importance of having a definite object in view.

Means of progress: Let us at once do something more for God than we are doing at present. Let us examine what we actually do and see what it amounts to, and how far it exacts any effort from us. And do not let us be hasty in deciding that we cannot afford to do more at present. Be cautious; but be generous as well.

There is something which we can infallibly do, and that is, put a more intense spirit into what we actually do; aim each of our actions to the greater glory of God, and inwardly unite our will to his in all we plan, or do, or suffer. Pray for a greater desire of perfection. It is in reality praying against worldliness, accustoming ourselves to unworldly standards and ideas.

RECOLLECTION.

Recollection is a double attention which we pay first to God and secondly to ourselves; and without vehemence or straining, yet not without some painful effort, it must be as unintermitting as possible. The necessity of it is so great that nothing in the whole of the spiritual life, love excepted, is more necessary. We cannot otherwise acquire the habit of walking constantly in the presence of God. The habit of recollection is only to be acquired by degrees. There is no royal road to it.

Until we feel the presence of God habitually, and can revert to him easily, it is astonishing with what readiness other subjects can preoccupy and engross us; and it is just this which we cannot afford to let them do. Newspapers keep not a few back from perfection.

The practice of retaining some spiritual flower, maxim, or resolution from our morning's meditation, in order to supply us with matter for ejaculatory prayer during the day, is a great help in acquiring recollection.

But the greatest help of all is to act slowly. Eagerness, anxiety, indeliberation, precipitancy—these are all fatal to recollection. Let us do everything leisurely, measuredly, slowly, and we shall soon become recollected. Nature likes to have much to do, and to run from one thing to another; and grace is just the opposite of this.

TEMPTATIONS.

Temptations are the raw material of glory; and the management of them is as great a work as the government of an empire, and requires a vigilance as incessant and as universal. In one sense, all temptations consist in an alliance between what is within us and what is without us.

Wherever temptation is, there God is also. There is not one which his will has not permitted, and there is not a permission which is not an act of love as well. The devil cannot lay a finger on the child until its loving Father has prescribed the exact conditions, and has forewarned the soul by his inspirations, and has forearmed it with proportionable succors of grace. The devil is simply our fellow-creature, and a conquered and blighted creature. He is continually overreaching himself.

Delectation is not consent. We are not the masters of the first indeliberate movements of our own hearts and minds. The enemy may run his hand flourishingly over the keys before we are aware. But there must be a deliberate acceptation and retention of the delectation before it can amount to consent or become a sin.

It is impossible for us to be altogether free from

distractions, useless to attempt it, and foolish to be dejected because we have not accomplished that impossibility. Conscious and deliberate acquiescence in and retention of distractions are, of course, our own affair; for it is in our power to withhold them; but the indeliberate occupation of our minds by them it is not in our power to prevent. Nothing can hinder bitter thoughts from disturbing us, wrong thoughts from staining us, and vain thoughts from disquieting and fatiguing us. The first sort of distractions are sand, the second pitch, and the third straw.

THE WORK OF PATIENCE.

Patience sanctifies for four reasons principally. The circumstances which exact its exercise come upon us from without; we have no control over them; they may come upon us at all moments; and they always involve the sacrifice or the mortification of our own will and way.

We may say that, partly from our own badness and partly from theirs, all mankind, far and near, kindred and strangers, are a trial to our patience in some way or other.

Almost every circumstance in life has a manner, time, place, and degree by which it tries our patience; and it is not too much to say, especially in the earlier stages of the devout life, that this exercise does more for us than fast or discipline, and that when we can go through with it for love of the sweetness of Jesus we are not far from interior holiness.

The English spirit of always standing up for our rights is fatal to perfection. It is the opposite of that charity of which the apostle says that it seeks not its own. Now this spirit is admirably mortified by the exercise of patience. It involves also a continual practice of the presence of God; for we may be come upon at any moment for an almost heroic display of good temper. And it is a short road to unselfishness; for nothing is left to self. All that seems to belong most intimately to self, to be self's private property, such as time, home, and rest, are invaded by these continual trials of patience. The family is full of such opportunities. It may be added, for it is no slight thing, that there is not a spiritual exercise less open to delusion than is this. If it is true of any one grace, besides charity, it is true of patience, that it is the beauty of holiness.

There is a vast difference between hatred of self and impatience with self. The more of the first we have the better, and the less of the last. Once let us surmount the difficulty of being patient with ourselves, and the road to perfection lies clear and unobstructed before us. But what do we mean by impatience with self? Fretting under temptations, and mistaking their real nature, and their real value also. In actual sin being more vexed at the lowering of our own self-esteem than being grieved at God's dishonor. In being surprised and irritated at our own want of self-control because of our subjection to unworthy habits. Being annoyed at our own want of sensible devotion, as if it were at all in our own power. To these symptoms we may add a sort of querulousness about the want of spiritual progress, as if we were to be saints in a month. These dangerous symptoms of impatience with self come from one or other of four causes: self-love, want of humility, the absence of a true estimate of the huge difficulties of the spiritual life, and an obstinate disinclination to walk by faith.

SIMPLICITY.

Simplicity aims at one end, seeks one object, is occupied with one work, and lives with singleness of heart. In its relations with God it puts away all multitude, all capriciousness, all distraction, all detachment, and its strength lies in its unity of purpose and its concentration of effort. In its relations with others it is gentle, open, without disguise, without insincerity, without flattery, and without deceit. There are hundreds of things which do not amount to lies, but which are contrary to the beautiful perfection of simplicity. There is a speech and a silence, there are looks, manners, permissions, concealments, dubious smiles, pretended inadvertencies, unworthy connivances, and intentional distractions, which grieve the Holy Spirit, and make sad ravages of the interior soul, though they are far short of absolute falsehood. If you would be perfect you must be true to a scruple. A hair's breadth of deceit must be to you as if it were a mile of positive untruth. Diplomacy of manner, way, and speech, circuitous routes for courtesy's sake, giving things the wrong names, and being silent when silence is really speech—these things are injuring men's sanctity, and causing saints to break in the mold, and frustrating beautiful purposes of grace every day. Christian simplicity, or holy truthfulness, requires, first, that we be truthful with ourselves; secondly, that we be truthful with others; and, thirdly, that we be truthful with God.

SELF-DECEIT.

It is the hardest thing in the world to acquire a knowledge of self. Are we really taking pains to do it? It is a sad annoyance when others find us out, for it mostly lowers their opinion of us; but the saddest annoyance of all to our poor nature is to find ourselves out; for if we lose self's good opinion we are forlorn indeed.

People are dishonest with themselves, either from the dislike of exertion, or from a suspicion that investigation will compel them to commit themselves to God or definitely deny him something, both of which they are equally anxious to avoid. There is hardly a man or woman in the world who has not got some corner of self into which he or she fears to venture with a light.

How very little do even good persons know themselves! Much of what they think is the work of grace about them is simply the providential accident of their circumstances. Self-love knows how to blend most skillfully its ideal with its realization of its ideal, so that not only shall nobody else know what is theory and what is practice, but even self shall not be able, at least with anything like assurance, to discern between the two.

There is no entanglement in creation like the entanglement of self-deceit; and there is this peculiarity about all its varieties, that they are all of them swift diseases, tending to become so very soon, and at such early stages, very difficult to cure. Its characteristic is deep-seated inveteracy. Self-deceit is very sore and sensitive when touched, though it is for the most part very hard to touch.

The higher operations of grace are more subject to delusion than the lower, except the very highest, which have to do with the soul's uttermost union with God. Very few even of those aiming at perfection rise above the middle graces. Hence it is practically the common rule that the higher men rise in the spiritual life the more subject they become to the insidious operations of self-deceit.

General simplicity of life is an antagonistic power to it. A man who habitually thinks of God, or one who thinks of God first and himself second, or one who does not sensibly live and act under the eyes and tongues of others, or one who does his duty lovingly, making few returns upon self, is as nearly an impossible subject for the greater triumphs of self-deceit as can be found among us poor, self-loving, self-seeking creatures.

The cure of self-deceit is not a thing which can be done once for all and then be over. It is a lifelong work. The first remedy is a great distrust of self, not merely in a general way, but in a very particular way. We must distrust ourselves precisely at the most privileged times and places, making it of faith to ourselves that when we are most sure we are in the right we are most surely in the wrong. Meditation on the attributes of God is another defense against self-deceit. The likeness of God is the aim of holiness, and we unconsciously imitate that which is a frequent subject of our meditation. The face of God will make us real. Communion with God eats away our unreality.

THE EVIL OF TAKING OFFENSE.

To give offense is a great fault, but to take offense is a greater fault. It implies a greater amount of wrongness in ourselves, and it does a greater amount of mischief to others. I do not remember to have read of any saint who ever took offense. The habit of taking offense implies a quiet pride which is altogether unconscious how proud it is. The habit of taking offense implies also a fund of uncharitableness deep down in us, which grace and interior mortification have not reached. Contemporaneously with the offense we have taken there has been some wounded feeling or other in an excited state within us. When we are in good humor we do not take offense.

Is it often allowable to judge our neighbor? Surely we know it to be the rarest thing possible. Yet we cannot take offense without, first, forming a judgment; secondly, forming an unfavorable judgment; thirdly, deliberately entertaining it as a motive power; and, fourthly, doing all this, for the most part, in the subject-matter of piety, which in nine cases out of ten our obvious ignorance withdraws from our jurisdiction.

A thoughtless or a shallow man is more likely to take offense than any other. He can conceive of nothing but what he sees upon the surface. He has but little self-knowledge, and hardly suspects the variety or complication of his own motives. Much less, then, is he likely to divine in a discerning way the hidden causes, the hidden excuses, the hidden temptations, which may lie, and always do lie, behind the actions of others.

Readiness to take offense is a great hindrance to the attainment of perfection. It hinders us in the acquisition of self-knowledge. No one is so blind to his own faults as a man who has the habit of detecting the faults of others. A man who is apt to take offense is never a blithe or a genial man. He is not made for happiness; and was ever a melancholy man made into a saint? A downcast man is raw material which can only be manufactured into a very ordinary Christian.

If it is not quite the same thing with censoriousness, who shall draw the line between them? Furthermore, it destroys our influence with others. We irritate where we ought to enliven. To be suspected of want of sympathy is to be disabled as an apostle. He who is critical will necessarily be unpersuasive.

In what does perfection consist? In a childlike, shortsighted charity which believes all things; in a grand, supernatural conviction that everyone is better than ourselves; in estimating far too low the amount of evil in the world; in looking far too exclusively on what is good; in the ingenuity of kind constructions; in our inattention, hardly intelligible, to the faults of others; in a graceful perversity of incredulousness about scandal or offenses. This is the temper and genius of saints and saintlike men. It is a radiant, energetic faith that man's slowness and coldness will not interfere with the success of God's glory. No shadow of moroseness ever falls over the bright mind of a saint. Now, is not all this the very opposite of the temper and spirit of a man who is apt to take offense? The difference is so plain that it is needless to comment on it. He is happy who on his dying bed can say, "No one has ever given me offense in my life." He has either not seen his neighbor's faults, or, when he saw them, the sight had to reach him through so much sunshine of his own that they did not strike him so much as faults to blame, but rather as reasons for a deeper and a tenderer love.

KINDNESS.

Kindness has converted more sinners than zeal, eloquence, or learning; and these three last have never converted anyone unless they were kind also.

Few men can do without praise, and there are few circumstances under which a man can be praised without injuring him. But kindness has all the virtues of praise without its vices. Praise always implies some degree of condescension, and condescension is a thing intrinsically ungraceful; whereas kindness is the most graceful attitude one man can assume toward another. So here is a most important work that kindness does. It supplies the place of praise.

Moreover, kindness is infectious. It makes others kind. The kindest men are generally those who have received the greatest number of kindnesses. A proud man is seldom a kind man. Humility makes us kind, and kindness makes us humble. It is the easiest road to humility, and infallible as well as easy. A kind man is a man who is never self-occupied. He is genial; he is sympathetic; he is brave.

Kind thoughts are rarer than either kind words or kind deeds. They imply a great deal of thinking about others. This in itself is rare. But they imply also a great deal of thinking about others without the thoughts being criticisms. This is rarer still. It seems to me that our thoughts are a more true measure of ourselves than our actions are. They are not under the control of human respect. It is not easy for them to be ashamed of themselves. They have no witnesses but God.

Kind thoughts, for the most part, imply a low opinion of self. They are an inward praise of others, and because inward, therefore genuine. The kind-thoughted man has no rights to defend, no self-importance to push. He finds others pleasanter to deal with than self; and others find him so pleasant to deal with that love follows him wherever he goes. Kind interpretations are imitations of the merciful ingenuity of the Creator finding excuses for his creatures. Have we not always found in our past experience that on the whole our kind interpretations were truer than our harsh ones?

A man is very much himself what he thinks of others. Even a well-founded suspicion more or less degrades a man. He is unavoidably the worse man in consequence of having entertained it. Virtue grows in us under the influence of kindly judgments, as if they were its nutriment. But in the case of harsh judgments we find we often fall into the sin of which we have judged another guilty.

Above all things the practice of kind thoughts is our main help to that complete government of the tongue which we all so much covet. The interior beauty of a soul through habitual kindliness of thought is greater than our words can tell. To such a man life is a perpetual bright evening, with all things calm, and fragrant, and restful.

Kind words are the music of the world. There is hardly a power on earth equal to them. It is by voice and words that men mesmerize each other. Happiness and kindness go together. The double reward of kind words is the happiness they cause in others and the happiness they cause in ourselves. Is there any happiness in the world like the happiness of a disposition made happy by the happiness of others? There is no joy to be compared with it.

We become kinder by saying kind words. A kind-worded man is a genial man; and geniality is power. Geniality is the best controversy. Satire will not convert men. Hell threatened very kindly is more persuasive than a biting truth about a man's false position.

We may put down clever speeches as the first and greatest difficulty in the way of kind words. A man who lays himself out to amuse is never a safe man to have for a friend or even for an acquaintance. He is not a man whom anyone really loves or respects. He is never innocent. He is forever jostling charity by the pungency of his criticisms, and wounding justice by his revelation of secrets.

The grass of the field is better than the cedars of Lebanon. It feeds more, and it rests the eye better—that thymy, daisy-eyed carpet, making earth sweet, and fair, and homelike. Kindness is the turf of the spiritual world, whereon the sheep of Christ feed quietly beneath the Shepherd's eye. Kindness is the occupation of the whole man by the atmosphere and spirit of heaven.

EDWARD MEYRICK GOULBURN.

IN barest outline the salient facts of Dr. Goulburn's life are as follows: Born 1818; educated at Eton and Oxford, where he graduated in 1839; ordained deacon in 1842, and priest in 1843; elected Fellow of Merton College, 1843; curate of Holywell, Oxford, 1841-50; head-master of Rugby School, 1850-58; prebendary of St. Paul's, 1858; one of her majesty's chaplains, and vicar of St. John's, Paddington, London, 1859-66; Dean of Norwich, 1866-89. He received the degree of D.C.L. in 1850, and of D.D. in 1856. He died May 3, 1897.

His writings have been very numerous, and very highly valued. The four from which the following extracts are taken are these: *Thoughts on Personal Religion, being a treatise on the Christian Life in its two chief elements, Devotion and Practice; The Pursuit of Holiness, a sequel to Thoughts on Personal Religion, being designed to carry the reader somewhat further onward in the Spiritual Life; An Introduction to the Devotional Study of the Holy Scriptures,* and *The Idle Word, short religious essays upon the Gift of Speech and its Employment in Conversation.* They have all passed through many editions (the first one about twenty), and because of their exceeding clearness and simplicity of style, as well as sound, sensible counsel, have been widely useful. Dean Goulburn is one of the very few in the present generation who by their solid services to the cause of religion deserve to be ranked with the great spiritual masters of the past, whom it is clear that he has closely studied.

DO ALL FOR GOD.

First, before you go forth to your daily task, establish your mind thoroughly in the truth, that all the lawful and necessary pursuits of the world are so many departments of God's great harvest field, in which he has called Christians to go forth and labor for him. Let us regard them all as, at least, if nothing more, wheels of the great world-system whose revolutions are bringing on the second advent and kingdom of Christ. Then, imagining yourself for a moment under no obligation to pursue your particular calling, undertake it with the deliberate and conscious intention of furthering his work and will. Choose it with your whole will as the path in which he would have you to follow him, and the task to which he has called you. Consecrate it to him by a few moments of secret prayer, imploring him to take it up with the great scheme of his service, and to make it all, humble, weak, and sinful as it is, instrumental in furthering his designs. Then put your hand to it bravely, endeavoring to keep before the mind the aim of pleasing him by diligence and zeal. Imagine Jesus examining your work as he will do at the last day, and strive that there may be no flaw in it; that it may be thoroughly well executed, both in its outer manner and inner spirit. At the beginning and end of every considerable action renew the holy intention of the morning.

DO ALL IN GOD.

Endeavor to make your heart a little sanctuary in which you may continually realize the presence of God, and from which unhallowed thoughts, and even vain thoughts, must carefully be excluded.

What we recommend, and what is surely attainable, is the mere consciousness that God's eye is upon us. Just as no speaker for a moment forgets, or can forget, that the eyes of his audience are upon him, and this does not interfere with the most active operations of mind, so with the presence of God. It is to be secured in the same way by which all other results in the spiritual life are obtained—by trustful, expectant, sanguine prayer and effort. We should call the attention definitely to God's presence, as occasion offers, at the necessary breaks or periods in our work, and occasionally mingle with the act of recollection two or three words of secret prayer which may suggest themselves. And it will be found in course of time that the constant recurrence of the thoughts to God will pass into an instinctive consciousness of his presence, and that the mind will acquire a tendency to gravitate toward him at all times, which will operate easily and naturally as soon as it is relieved of the strain which worldly affairs put upon it.

INTERRUPTIONS.

Are you a firm believer in the providence of God? Do you believe that the whole of your affairs—trivial as well as great, irregular as well as in the ordinary course—are under his absolute, daily, hourly supervision and control? that nothing can possibly arise to you or any other which is not foreseen by him, brought by him within the circle of his great plan? that the little incidents of each day, as well as the solemn crises of life, are his ordering? Then you admit that the occurrences of each day, however unlooked-for, however contrary to expectation, are God-sent, and those which affect *you* sent specially and with discrimination to yourself.

There is many a man who says, "I will conform myself to the general indications of God's will made to me by his word;" comparatively few who say, "I will conform myself to the special indications of God's will made to me by his providence." Why so few?

Here then lies the real remedy for the uneasiness of mind which is caused by interruptions. View them as part of God's loving and wise plan for your day, and try to make out his meaning in sending them. When in your hour of morning devotion you distribute your time beforehand (as it is in every way wise and proper to do) let it always be with the proviso that the said arrangement shall be subject to modifications by God's plan for you as that plan shall unfold itself hour by hour to your apprehensions. The radical fault of our nature, be it remembered, is self-will; and we little suspect how largely self-will and self-pleasing may be at the bottom of plans and pursuits which still have God's glory and the furtherance of his service for their professed end.

Suppose the mind to be well grounded in the truth that God's foresight and forearrangement embrace all which seems to us an interruption—that in this interruption lies awaiting us a good work in which it is part of his eternal counsel that we should walk, or a good frame of mind which he wishes us to cultivate; then we are forearmed against surprises and contradictions; we have found an alchemy which converts each unforeseen and untoward occurrence into gold; and the balm of peace distills upon our heart, though we be disappointed of the end which we had proposed to ourselves. Let us seek to grasp the true notion of providence; for in it there is peace and deep repose of soul.

PURITY OF INTENTION.

Perfect purity of intention is the highest spiritual state, a state which probably the holiest man has never reached, but to which all real children of God are in different measures approximating. Our defectiveness of intention should be, and may be, by self-examination, and careful attention and prayer, remedied. Let the motives as well as the actions be scrutinized in self-examination. Ask: "Should I have done this, or done it with equal zeal, had no eye of man been upon me? Should I have resisted this temptation if there had been no check upon me from human law or public opinion? Should I have acted thus faithfully and conscientiously without the stimulus of human praise?" Let us cultivate particularly, and strive to acquit ourselves well in, those actions of the Christian life which are in their nature private, and cannot come abroad. For example, private prayer and private study of the Scriptures. Exercises such as these are more or less a satisfactory test of religious character, because they are incapable of being prompted by human respect. And we may apply the same remark to all the ordinary actions and commonplace business of life, which must be transacted by all in the same way, and may be transacted by the Christian with a spiritual intention. What does growing in grace mean but that this spiritual intention should lengthen its reach—should extend itself more and more to every corner of our life?

The meeting all calls upon us, however humble, with the thought that they come to us in the way of God's providence, and are indications of the quarter in which he would have us direct our energies, is a great means of purifying our intention, and so of advancing in spirituality. For nobody is aware what is going on in our hearts when we meet these calls in a devout spirit; our friends only see us doing commonplace things, which others do, and give us no credit. But in meeting such calls we have praise of God, who, like a good Father, marks with a smile of approbation the humblest efforts of his children to please him.

To live holily is nothing else than, in everything we do, to act from a single desire to please God out of love to him, and from no other aim whatever.

HATRED OF EVIL.

By way of testing the affections of our hearts toward God let us ascertain how we are disposed toward his opposite—evil. To hate evil is something far more than merely to shun or avoid it. If we do not hate impurity, sicken at the sight and thought of it, and turn away with disgust, it is out of the question that we can love God, who is purity. It is quite possible not to be implicated personally in sin and yet to treat it, when witnessed or heard of in others, with levity and indifference. There can be no question that, if a man were in a perfect moral state, moral evil would affect his mind as sensibly, and in as lively a manner—would, in short, be as much of an affliction to him as pain is to his physical frame. Our Lord Jesus Christ not only loathed the grosser forms of evil, but he flung from him with abhorrence every unspiritual suggestion, such as that once made to him by the apostle Peter, to decline the cross and consult his own ease.

Love, as a Christian grace, is an altogether different thing from many qualities which usurp its name. A different thing from that easy pliability of will which is called good nature, but which in fact resolves itself into indolence and languor of character. On the contrary, in all real love there is strength, strength of will and strength of character. In all real love there is wrapped up hatred against that evil which counteracts goodness. Generally speaking, the truest Christians have in them the greatest force of character. There must be resoluteness to obtain the prize. The salt of decision and energy must be mixed with the oil of love. Again, Christian love is a very different thing from that indifference to theological error which, in these latitudinarian days, too often apes its manners and mimics its phraseology. In lesser (or doubtful) points, not affecting the vitality of God's truth, our maxim must be tolerance to the very utmost; nay (more than tolerance), a catholic acknowledgment of whatever is good and wise in other Christian Churches. But where the error mutilates the vital parts of the truth, there love can only appear in its form of hatred of evil. It is a very serious breach of love to pay compliments to false doctrine. Our blessed Lord and his apostles never did so.

LOVE OF OUR NEIGHBOR.

What we are required to love in our neighbor is the image of God in him. Every soul has a fragment of this image in its lowest depth, though it may be overlaid by all manner of rubbish—infirmity, imperfection, frivolity, sin. The true Christian studies the happy art of making the most of everyone with whom he is thrown in contact—of recognizing in each soul and of eliciting from it that feature of heart and mind in which stands the relationship of that particular soul to God. It is this true self of our neighbor that we are required to love. We are not required to love infirmities or imperfections; nay, we could not do so if required; for infirmities and imperfections are naturally repelling. God must hate sin in its every form; between him and insincerity, untruthfulness, peevishness, petulance, ill-temper—above all, perhaps, between him and selfishness—there must be an eternal antipathy. And yet nothing is more certain than that, while God hates my selfishness and untruthfulness, he deeply and tenderly loves *me* with an individualizing love. And he would have me love my neighbor exactly as he loves me; fastening my regard upon his true self, upon the feature of God's image which is reflected in his soul, and bearing with his infirmities out of this esteem for the true self.

Our love of our neighbor must be brought to practical tests. Are we doing anything to help him? making sacrifices for him, of money, or time, or pleasures? It is an excellent spiritual precept, whenever a good desire springs up in our heart, to stereotype and make it permanent; in other words, to bring the good desire to good effect by an effort in that direction. Secondly, our prayers for others furnish a good practical test as to the genuineness of the love we bear them. What approach are we making to the great model of the Lord's Prayer, which does not contain any petition exclusively directed to our own wants? Do we pray for others at all? And, if we do, is this exercise considered by us merely as an ornamental appendage to our other prayers, but as in no wise essential to their acceptability with God? Seek to make your prayers for others specific, so far as your knowledge of their character and circumstances allows. Pray for them sympathetically. And pray for this sympathy, while you endeavor, by careful consideration of their case, to excite it within yourself. Our efforts for others, whether of prayer or benevolence, are not lost. If *they* are not benefited by them *we* are: in increase of light, and power, and comfort, in whispers of mercy and peace, they return again into our bosom.

HOW TO WORK.

Do your work under the eye of the heavenly Master, and look up in his face from time to time for his help and blessing; an internal colloquy with him ever and anon, so far from being a distraction, will be a furtherance. For no work can in any high sense prosper which is not done in a bright, elastic spirit; and there is no means of keeping the spirit bright and elastic but by keeping it near to God. Another point is, never to allow ourselves to think of our work as a distraction or a hindrance to piety. Regard it in its true light morally and spiritually.

But the most important point of advice in an age like ours, when men in all conditions of life are overweighted with work, and in a country like ours, whose inhabitants are so little meditative and so constitutionally busy, is to aim rather at doing well what we do than at getting through much. Francis of Sales thought that the great bane of the spiritual life in most men is that eagerness and undue activity of the natural mind which leads to precipitancy and hurry. The remedy is to recommend the work to God and humbly ask his blessing and his aid, as we may do with the utmost confidence if the work be really that which his providence has assigned to us; then, resolutely to refuse to attend to more than one thing at a time, and to let everything else drop till that one thing is done. Other things must wait. Some of them we shall be able probably to do by and by. Not a few of them will do themselves. And some of them, may be, we shall have to leave undone. Let us not be disquieted. If the spirit of the doer have been right all will be well. If we could do our work in a brighter and less anxious spirit it would wear us less. It is worry, not work, that wears.

FAITH IN GOD.

What is faith? It may be defined as the faculty by which we realize unseen things, the faculty of spiritual touch. Faith is the only faculty which grasps the unseen, which brings it home to us and gives it a living power, so that we have such a conviction of its reality as to live under its influence.

When directed toward God or Christ faith takes the form of trust. But how can we trust a person without a high conception of his character? Seek, then, to feed and nourish in your mind great conceptions of him with whom you have to do. Expand and exalt your notions of him by every means in your power. Large and exalted conceptions of God are the spring of all virtue.

We may know him in part from his creation. Consider the lilies of the field, and the fowls of the air. Why, because superior edification and clearer light are to be had from our own Bible, are we to look down upon the edification and light which are to be derived from the Bible of the Gentiles? Might we not on the same principle neglect the Old Testament because the New is of superior importance?

The life of true religion, then, is an experimental knowledge, a heart-knowledge, of God—such a thorough appreciation of the excellence and beauty of his character as really contents and satisfies the soul, even when earthly sources of happiness fail. The knowledge of God is gained, as the knowledge of man is gained, by living much with him. If we only come across a man occasionally, and in public, and see nothing of him in his private and domestic life, we cannot be said to know him. All the knowledge of God which many professing Christians have is derived from a formal salute which they make to him in their prayers, when they rise up in the morning and lie down at night. While this state of things lasts no great progress in the Christian life can possibly be made. No progress would be made even if they were to offer stated prayers seven times a day instead of twice. But try to draw down God into your daily work; consult him about it; offer it to him as a contribution to his service; ask him to help you in it; ask him to bless it; do it as to the Lord, and not unto men; refer to him in your temptations; go back *at once* to his bosom when you are conscious of a departure from him; in short, walk hand in hand with God through life, dreading above all things to quit his side, and assured that as soon as you do so you will fall into mischief and trouble; seek not so much to pray as to live in an *atmosphere* of prayer, lifting up your heart momentarily to him in varied expressions of devotion as the various occasions of life may prompt, adoring him, thanking him, resigning your will to him many times a day, and more or less all day; and you shall thus, as you advance in this practice, as it becomes more and more habitual to you, increase in that knowledge of God which fully contents and satisfies the soul.

Again, it is obvious that the knowledge of God of which we speak may be obtained from studying his mind as it is given us in the Holy Scriptures. We may be said to know an author when we have so carefully and constantly read his works as to imbibe his spirit. There is a study of Scripture which is analogous to ejaculatory prayer—not a stated study (though of course the stated study of it may not be neglected), but a study which inweaves the Word into the daily life of the Christian, a rumination which can be carried on without book, and which is more or less continual.

Again, the discipline of life will very much contribute toward the knowledge of God. Those who desire to have a practical and experimental, as distinct from a speculative, knowledge of him will study him in these his dealings; they will try to discern the lesson of every part of their own experience, if haply it may teach them something of him with whom they have to do, and will thus have his wisdom, power, and love impressed upon them in a way in which nothing short of experience can impress.

THE MORTIFICATION OF OUR MEMBERS.

First, it should be deeply considered what it is that has to be mortified in us—that it is the affection to created good, not in one particular shape, but in all its forms. The first step, therefore, to be taken by him who would exercise a wise mortification is to consider deeply in what form or forms of earthly good he is naturally disposed to place his happiness; what forms yield him, constituted as he is, most comfort, most gratification. Whatever it be—human esteem, luxurious ease, sympathy, the gratification of ambition, amusement—there let him exercise a jealous watchfulness over himself; there let him mortify his will. To mortify the will is often a far greater cross than to inflict the severest penance on the body. There let him lay by force restrictions upon himself, sometimes sharply refusing all indulgence to the propensity, however in itself innocent, never at any time giving it too free a rein. The more intensely a man realizes unseen and eternal things the more he can afford to dispense with the things that are seen and are temporal. Mortification is not an end in itself, it is but a means to an end—that end being the springing up in our hearts of a fountain of eternal joy. And therefore to cultivate a taste for spiritual enjoyment, and to place one's contentment and satisfaction more and more exclusively in the contemplation of God and in communion with him, is the way to grow in the spirit of mortification, without which spirit the bare acts of it have little or no value.

EMOTION AND AFFECTION.

The true life of the soul is in its affections, not in its emotions. Emotions are impossible (according to the law of our minds) except at a crisis and moment of convulsion. And he who seeks for them under ordinary circumstances will run the risk of making his religion morbid. There are two safe signs, in our normal spiritual life, that we love Christ. One is *confidence*. The habit of exposing the contents of the heart to Christ, of referring all our actions to his will, of commending all our troubles to his care, and all our difficulties to his direction; the realizing him as being by our side, always sympathizing, always inviting our confidence, always ready and willing to help us; the being sincere in all our dealings with him, and perfectly single-minded in seeking to know his will—this is one great test of love for him, which, if really found in us in a small degree, is worth a large amount of high-flown feeling.

And the second test is that we seek to please him.

To attempt to please Christ is not only to act in compliance with the general indications of his will which are made to us in his Word, but to be on the watch for opportunities of doing him service, and to embrace those opportunities whenever they arise; it is to be guided by his eye, as well as by the express directions of his voice, and to find in the sense of his favor and approving smile the strongest stimulus to duty. Whoever feels and acts thus toward him must love him, however little of sensible emotion he may experience. Emotion may be defined as affection quickened by a crisis. But then it is not at all essential to the existence or genuineness of an affection that it should be thus quickened.

The will is the sphere in which all genuine love for Christ displays itself. "If ye love me, keep my commandments," Christ says. Your love for me must be an affection of the will; it must be a moral choice of me in preference to sin and the world, and must show itself in embracing my will both by active obedience and passive submission. It must be grounded upon a perception of my excellence and of the benefits received from me, and must enable you to find in the single-minded effort to please me a satisfaction purer, higher than is to be found in any earthly gratifications, and of a different order.

Reader, how far does your love for Christ reside in the will, in the judgment and moral sense? Do you live much with him, and love to live with him, in thought and in prayer? Do you honor him by drawing him into use in all his offices of grace? Can you yield up your will into his hands, to choose for yourself nothing else than he chooses for you? Does the satisfaction of trying to please him excel every other in a certain high and pure flavor? These are the questions which must determine the genuineness of our love for him. And genuine love is the only safe evidence of genuine faith in him. And on faith in him is suspended our salvation.

WHAT SHUTS CHRIST FROM US?

What is it which occupies the room in our hearts which he seeks? Two things principally, under which all others will fall: first, self-will, and then confidence in the creature for happiness.

The least trace of self-will excludes *pro tanto* God and his working from the soul. Absolute surrender to his will and word in everything is the only condition on which the Lord will take up his abode in the depth of the soul, and give to the heart that calm and repose which only his presence can give. There is, alas! many a will which does not sit loose upon its pivot, but is fixed in the quarter to which its natural inclinations point, and which moves not, therefore, when the breath of God's Spirit seeks to turn it. There are many Christians who have not that delicate sensibility to God's inspirations which he loves to find in a soul, and which, when he does find it, enables him to do many mighty works therein.

As a man increases in earnest love to Christ, a delicate tact grows up within him, a spiritual instinct, which teaches him (without any book) what he ought to say and do, and what he had better avoid on each particular occasion. God's children know the meaning of his eye. They know, by the glance he gives them, what path he would have them pursue, and what avoid. He never leaves them without an interior indication of his will, if they have but one desire, that of pleasing him. And why these indications are so rarely made is that God sees people are not quite disposed to accept them, not prepared in all things to move in the direction indicated. The soul must be empty of self-will before God can work in it.

Confidence in the creature for happiness. Who shall say (without very special grace and an extraordinary measure of divine illumination) how far his affection is set upon the earthly blessings with which his cup is crowned? It is but too easy to deceive ourselves in this matter while the earthly blessings remain with us. If God sees the affections of trust and love twining too closely around the creature, in very faithfulness to us he must tear them away, and cause a painful bleeding of the heart. The only way to keep our earthly treasures, on the assumption that we are God's true people, is, while we thankfully hold them of God, to mortify all undue attachment to them, and sternly to refuse to idolize them.

There can be no blessing without a risk. How long it is before a soul can perfectly unlearn trust in the creatures! Does it ever completely unlearn this trust, while life lasts, and while the body of sin and death clogs it? I suppose not. We must learn the art of tasting the various blessings with which God crowns our cup without being engrossed or taken up with them, without suffering them to quench the high aspirations of our soul after communion with God. This is a lesson which it takes long practice, much self-control, and great discipline of God's providence and Spirit to teach.

PEACE OF MIND AND HEART.

Peace is a very sensitive guest, apt to take flight at the slightest affront. We shall never know what it is to live in peace until we know what it is to live thoroughly in the present, rather than in the past or the future. We must restore to their right places and functions the acquiescence and the forward impulse which there are in our nature; be easily satisfied as regards our condition, so as not to indulge a wish for the change of it; be deeply dissatisfied with the little we know of God and of ourselves, and the miserably little we do for him. Let our whole care be to serve God in the present moment of our lives, being anxious for nothing. Deal with a fruitless anxiety just as you would deal with an impure or a resentful motion of the heart. Shut the door on it at once, and with one or two short ejaculatory prayers rouse the will and turn the thoughts in a different direction. Having made known your wishes to God in prayer, and begged him to deal in the matter, not according to your shortsighted views, but as seems best to his wisdom and love, leave it with him. If prudence and caution dictate that anything should be done to avert the evil you anticipate, or bring the blessing you desire, do it, and then think no more of the subject. Fruitless thinking is just so much waste of that mental and spiritual energy every atom of which you need for your spiritual progress. It is also a positive breach of God's precept, "Be anxious for nothing."

Try to realize God's presence; the realizing it ever so little has a wonderfully soothing and calming influence on the heart. Say secretly: "The Lord is in his holy temple (his temple of the inner man); keep silence, O my heart, before him." The mind wants steadying and setting right many times a day. It resembles a compass placed on a rickety table; the least stir of the table makes the needle swing round and point untrue. Let it settle, then, till it points aright. Be perfectly silent for a few moments, thinking of Jesus; there is an almost divine force in silence. Drop the thing that worries, that excites, that interests, that thwarts you; let it fall, like a sediment, to the bottom, until the soul is no longer turbid, that you may serve God with a quiet mind. We cannot serve him with a turbid one; it is a mere impossibility. The Spirit cannot make communications to a soul in a turbulent state, stormy with passion, rocked by anxiety, or fevered with indignation. Not until the wind, the earthquake, and the fire have subsided can God's still small voice be heard communing with man in the depths of his soul. Thus composing ourselves from time to time, setting the mind's needle true, we shall little by little approximate toward that devout frame which binds the soul to its true center, even while it travels through worldly business, worldly excitements, worldly cares.

DEVOTIONAL USE OF SCRIPTURE.

The established ordinance through which God addresses man is the Holy Scripture. Its general character evinces the necessity of meditation for those who desire to use it aright. The Scripture is rather a book of principles than of rules, of examples than of precepts; it is essentially an unsystematic book. Hence the right use of it must involve effort and exertion. From the examples a moral must be drawn. Never read the Scripture narrative without asking yourself what practical lessons are to be derived from it. From the rule, where a rule is given, the reader must apply himself to gather the principle. And, again, rules must be framed from principles. As men are in general constituted, rules, specific, strict, and stringent, are absolutely necessary to progress in the spiritual life.

Meditation on Scripture need not be limited to set times, but may be carried on profitably in any hour of solitude, and whenever the mind is not otherwise engaged. Possibly at some interval during the day you may be alone. Have recourse then to the passage of Scripture which you have previously lodged in your mind, and ask yourself seriously, as in the sight of God, what practical lessons it is designed to teach, what bearing it has upon your spiritual welfare. At first you will find it difficult to prevent the thoughts from flying off to other topics. The power of fixing the mind is only to be gained by habit. Perhaps a little effort of the fancy may here lend us some assistance. During a solitary walk, or at any other period of leisure, imagine that, when you return, you will be called upon to address an audience on the subject which you propose for meditation. It wonderfully disentangles all difficulties to consider how we could make plain to other minds the truth which is thus beset to our own.

The plan of a meditation on Holy Scripture: First, endeavor to realize the presence of God according to that conception of this great truth which best suits your own mind. Feel that he is here.

Secondly, call upon God as an essential condition of success, to inspire you with holy thoughts, and to bless them to your spiritual profit and growth in grace for Christ's sake. Do it very briefly, but with great earnestness. Thirdly, open the passage of Scripture which is to form the subject of meditation; or repeat it mentally. Fourthly, the Bible (in the original, if you know the language sufficiently well to make it available) being opened at the passage, picture to yourself the circumstances by an effort of the imagination. Fifthly, the circumstances having been pictured, next comes the exercise of the understanding upon the words. We reflect upon them, turn them over in our mind, endeavor to make out what they teach, what doctrine is wrapped up in them, and what duty. Sixthly, next follows the exercise of the affections and the will, incomparably the most important part of the whole meditation. In this consists the practical application of the little sermon to your own heart, in the absence of which it is useless, or in some respects worse than useless. It will be a good plan to allow any feeling which stirs within you, as you regard the truths of the passage, to express itself in prayer. Conclude all by an exercise of the will, that is, by one or more resolutions. It has been recommended also, before quitting the subject altogether, to pick out some one sentiment which has pleased us most, and to charge the memory with it during the remainder of the day, so that it may continually be recalled to mind at intervals, and be like a fragrant flower plucked from the garden and worn in the girdle, whose odor refreshes us amid the dust and turmoil of life. This last is the precept of the devout Francis de Sales, whose method of meditation we have followed.

PROPER FUNCTION OF WORDS.

What is the proper function of words, the end for which they were given, by fulfilling which they become good and escape the censure of being idle words? The first, and perhaps the lowest, end of words is to carry on the business of life. The second end is to refresh and entertain the mind. The world's wisest men have mingled mirth with earnestness; they have not gone about with starched visage, prim manner, or puritanical grimace. By way of preserving pure this offspring of the heart's merriment three cautions should be rigidly observed: First, from all our pleasantry must be banished any, even the remotest, allusion to impurity, which forms the staple of much of this world's wit. Secondly, all such sarcasms as hurt another person, wound his feelings, and give him unnecessary pain, are absolutely forbidden by the law of Christian love. Thirdly, all such pleasantries as bring anything sacred into ridicule—or, without bringing it actually into ridicule, connect with it, in the minds of others, ludicrous associations, so that they can never see the object or hear the words without the ludicrous observation being presented to them—are carefully to be eschewed.

A desire of gaining instruction is one of the first dispositions with which we must engage in conversation, if we desire to make it profitable, nay, even entertaining, to both parties. Let it be considered a fixed and ascertained truth that your neighbor, however he may be inferior to you in some points of station and attainment, is able to impart to you some information which you do not possess. This is not a fancy; it is a real truth. Let us, therefore, when either casually or by design we enter into company, set ourselves to the finding out what that something is. Make an effort to extract, from those with whom the occasions of life bring you into contact, that portion of useful knowledge which out of the common stock they have appropriated to themselves.

Idle words are forbidden by the Saviour. But by this he means useless words, conducive neither to instruction nor to innocent entertainment—words having no salt of wit or wisdom in them; flat, stale, dull, and unprofitable; thrown out to while away the time, to fill up a spare five minutes; words that are not consecrated by any seriousness of purpose whatever, that contribute nothing either to the carrying on of the necessary business of life, or harmless amusement, or to the lower or higher forms of instruction, or to the glory of Almighty God.

It is every man's duty, as it ought to be esteemed every man's privilege, to say a word for God in society wherever such a word may be discreetly and properly introduced; to be faithful with his more intimate friends in representing their defects of character and conduct; to be thankful himself for receiving such representations; and ever to be on the watch to arrest an opportunity for profitable conversation.

MISCELLANEOUS COUNSELS.

The greatest saints who ever lived, whether under the old or new dispensation, are on a level which is quite within our reach. The same forces of the spiritual world which were at their command, and the exertion of which made them such spiritual heroes, are open to us also. Why should we not follow them, even as they followed God and Christ? The reason is not to be sought in any disadvantages under which we labor in comparison with them. It is not that holiness was originally more congenial to their nature than to ours. It can be nothing but that laggardness of will, that indifference to high moral aims, that want of spiritual energy, that cheerful acquiescence in the popular standard of religion which has caused many a soul when "weighed in the balances" to be "found wanting," to be counted unworthy of the calling and the kingdom of God.

If we would bestow our efforts in the spiritual

life well and wisely we need not so much seek to do something religious as to do ordinary things in a religious manner, cultivating high and loving thoughts of God while we do our work, and seeking to do it well, where no eyes are upon us, from the view of pleasing him; and in all services to our fellow-men thinking of the image of God which lies hidden and overlaid with rubbish in their souls, as in ours, and of the enormous price of Christ's blood, which was paid down for all, showing how high must have been God's estimation of each of them. We shall never regret any amount of pains taken in doing common things as unto the Lord, and in striving to evince love to him by means of them.

The great question is whether, after every fall, the will recovers its spring and elasticity, and makes a fresh start with new and more fervent prayer and resolve. In order to any great attainment in the spiritual life there must be an indomitable resolve to try and try again, and still to begin anew amidst much failure and discouragement. It is by a constant series of new starts that the spiritual life is carried on.

To be right in the practical department of the Christian life is summed up in these three things: to work devoutly, to fight manfully, and to suffer patiently.

Resolve to know much of the inward life of religion. Cultivate in every possible way a spirit of private devotion. Determine to know the power of prayer as distinct from its form. Practice more and more in all companies and under all circumstances the thought of the presence of God. Seek more and more to throw a spiritual meaning and significance into your pursuit; to do it more simply and exclusively from the motive of pleasing God, and less from all other motives. Try, by a holy intention, to give even to the more trifling actions of the day a religious value. Let self occupy as little as possible of your thoughts. Care much for God's approval, and comparatively little for the impression you are making on others. Thus you will feed the inward light with oil, and it will shine.

Specific resolutions are of the greatest service in the spiritual life. They must be framed upon the knowledge of our weak points and besetting sins; and it is well every morning to draw up one or more of them after a foresight of the temptations to which we are liable to be exposed. Nothing is so likely to destroy that recollectedness of mind which is the very atmosphere of the spiritual life as unexpected incidents for which we are in no wise prepared, and which often stir in us sudden impulses of almost uncontrollable feeling. Let us arm ourselves for them, so far as possible, by a holy resolution, which will take its shape from the peculiar nature of the temptation offered—a resolution, perhaps, to busy ourselves in some useful work, and so divert the mind, or to give a soft answer which turns away wrath, or to repeat secretly a verse of some favorite hymn, or only to cast a mental glance on Christ crucified, which indeed is the most sovereign remedy against temptation known in the spiritual world.

If the Christian in every part of his active work for God sets God before him; if he is very jealous of the purity of his motives; if he is diligent in ejaculatory prayer; if, even in the little crosses and annoyances of the day, he regards the will of God who sends them, and takes them accordingly with sweetness and buoyancy of spirit; if he cultivates the habit of allowing the objects of nature and passing events to remind him of spiritual truth, and lead his mind upward; if, in short, he turns each incident of life into a spiritual exercise, and extracts from each a spiritual good—then he is cultivating the internal life, while he engages in the external; and while, on the one hand, he is expending the oil of grace, he is, on the other, laying in a fresh stock of it in his oil-vessels.

A DOZEN WORTHIES.

BESIDES the eight writers to each of whom, for reasons that seemed to us sufficient, we have given considerable space there are, of course, very many others whose stores of instructive counsel or experience might readily be drawn upon to any extent. But we have thought best to limit our further extracts to twelve authors, all of them loved and prized by multitudes and deserving to be introduced to those not yet familiar with their fame. Space permits us, much to our regret, to afford in each case only a taste of the quality of the volume in question, but these tastes will be in themselves helpful, and will serve the additional purpose of making many acquainted with good books to which their attention might not otherwise be drawn. We shall present them in the order of their age. Hence will come first

"THE CONFESSIONS OF ST. AUGUSTINE."

This is not the place for a delineation either of the life or the writings of this great man, so prominent among the fathers of the Church. Born at Tagaste, in Numidia, November 13, 354, he died at Hippo, in northern Africa, of which place he had been for thirty-five years bishop, August 28, 430. He wrote the *Confessions* in the year 398, eleven years after his baptism and three years after his consecration as bishop. There is a charm and simplicity in the style rarely, if ever, surpassed, which has endeared the little book to great numbers. It affords also a pleasing insight into the kind of life common in that far-distant period. But since the treatise relates almost wholly to the experiences of the author previous to conversion, his struggles with Manichæan error and licentious vice before his deliverance from these toils of Satan, it is not, as a whole, very profitable for ordinary perusal. His object apparently was to illustrate the goodness and forbearance of God in bringing him, despite manifold mistakes, infirmities, and sins, to a blessed haven of rest, that others might be strengthened against despair. Many translations of it from the original Latin have been made into various modern tongues, and many editions have been issued both in separate form and in connection with the other works.

The two sentences most frequently met in quotation from the *Confessions* are these: "Thou madest us for thyself, and our heart is restless until it rest in thee." "Give what thou enjoinest, and enjoin what thou wilt; for too little doth he love thee who loves anything with thee which he loveth not for thee."

The account which he gives of his conversion is exceedingly beautiful, and with this extract we must now be contented:

"Thou, Lord, didst turn me round toward myself, taking me from behind my back where I had placed me, unwilling to observe myself, and setting me before my face that I might see how foul I was, how crooked and defiled, bespotted and ulcerous. And I beheld and stood aghast; and whither to flee from myself I found not. . . . But when a deep consideration had from the secret bottom of my soul drawn together and heaped up all my misery in the sight of my heart, there arose a mighty storm, bringing a shower of tears. . . . I cast myself down, I know not how, under a certain fig tree, giving full vent to my tears; and the floods of mine eyes gushed out an acceptable sacrifice to thee. I sent up these sorrowful words: How long? how long, 'to-morrow and to-morrow?' Why not now? why not this hour an end to my uncleanness? So was I speaking and weeping in the most bitter contrition of my heart, when, lo! I heard from a neighboring house a voice, as of boy or girl, I know not, chanting and oft repeating, 'Take up and read; take up and read' (*Tolle, lege*). Instantly my countenance altered, I began to think most intently whether children were wont in any kind of play to sing such words; nor could I remember ever to have heard the like. So checking the torrent of my tears I arose, interpreting it to be no other than a command from God to open the book and read the first chapter I should find. I seized the volume, opened, and in silence read that section on which my eyes first fell: 'Not in rioting and drunkenness, not in chambering and wantonness, not in strife and envying: but put ye on the Lord Jesus Christ, and make not provision for the flesh' (Rom. xiii, 13, 14) in concupiscence. No further would I read; nor needed I. For instantly at the end of this sentence, by a light as it were of serenity infused into my heart, all the darkness of doubt vanished away. . . . Thence I go in to my mother; I tell her; she rejoiceth. I relate in order how it took place; she leaps for joy, and triumpheth, and blesseth thee."

WORKS OF JOHN TAULER.

Tauler was born at Strasburg about 1290, and died there June 16, 1361. He was the greatest preacher of his age, but it is not in that his main distinction lies. It is in his exceptional religious experience, and his connection with that remarkable band of Christian men known as "Friends of God." This was an extensive but slightly organized brotherhood, scattered over the upper provinces of the Rhine country, composed of those who sought for intimate communion with heaven, and in the midst of the abounding iniquities of the times held themselves to a high standard of personal piety. They laid great stress upon disinterested love, self-renunciation, and a constant loving fellowship both with one another and with the Holy Spirit.

Tauler was of honorable family and early devoted

to the priestly office. At the age of eighteen he became a Dominican monk and went to Paris to study theology. Returning to Strasburg, he began to preach with considerable success, but his sermons were not pervaded with the power which comes from a personal union with Christ. He was not brought into full freedom until more than fifty years of age. The instrument of his deliverance from the bondage of the law, from a too formal piety somewhat tainted with self-righteousness, was an uneducated layman named Nicholas, very many years his junior, but well taught in the things of the Spirit. Coming from Basel to Strasburg to hear the distinguished preacher, he speedily detected the lack in his experience, and was enabled to lead him on to much greater heights. When Tauler had once been brought, after two years' struggles, to see himself and his Saviour in the true light, the change in his sermons was immediate and great. The first time that he opened his mouth in public fourteen persons fell as if dead under the power of the word, and nearly thirty others were so deeply moved that they remained sitting in the churchyard long after the congregation was dismissed, unwilling to move away. A great revival began, both among those previously religious and among the worldly, a revival whose influence in Germany was widespread and far-reaching, reaching indeed in some of its effects down to the present day. The discourses which his disciples preserved had a decided influence upon Luther, who was accustomed to recommend them as the best sermons to be found in the German language.

For eighteen years after what may be called, perhaps, his second conversion, Tauler made progress in the divine life, rising to the place of highest esteem among his brethren and being accounted the holiest of God's children on earth. Men came from all quarters to consult him, and his usefulness continually extended. Nor did he lack for persecution, that supreme testimonial to goodness. He is every way worthy of largest honor and closest study. As one has well said: "No idle contemplation or passive asceticism finds the approval of Tauler, but a life of active love and pity, of patience and meekness—a life in the imitation of Christ. Tauler did not contradict the doctrines of his Church, but he was animated by an exalted reformatory spirit; his mysticism displayed a free, practical, evangelical tendency which has given it historical importance; and we may appropriately retain for him the title, early bestowed, of *Doctor Illuminatus.*"

The first collected edition of his sermons was printed at Leipsic in 1498, and very many others have followed. An English translation by Miss Winkworth was published at London in 1857, and an American reprint, edited by Dr. Hitchcock, was issued at New York in 1858. We shall have to confine our selection from his writings to the well-known *Discourse of Dr. Tauler with a Beggar,* which has been often quoted during these five centuries, and can never be read without profit:

"There was once a learned man who longed and prayed full eight years that God would show him some one to teach him the way of truth. And, on a time, as he was in a great longing, it was said unto him, 'Go to such a church porch, and there wilt thou find a man that shall show thee the way to blessedness.' So thither he went, and found there a poor man, whose feet were torn and covered with dust and dirt, and all his apparel scarce three hellers' [farthings] worth. He courteously saluted him, saying, 'God give you a good morning.'

"To which the beggar replied, 'I never remember to have had a bad morning.'

" 'God prosper you,' said the Doctor.

" 'What say you?' answered the beggar. 'I never was otherwise than prosperous.'

" 'I wish you all happiness,' replied the Doctor; 'but what do you mean by speaking in this manner?'

" 'Why,' said the poor man, 'I never was unhappy.'

" 'God bless you,' said the Doctor; 'explain yourself, for I cannot well understand your meaning.'

" 'Willingly,' quoth the poor man. 'You wished me a good morning, and I answered that I never had a bad morning; for if I am hungry I praise God; if I suffer cold I praise God; if it hail, snow, or rain, if the weather be fair or foul, I give praise to God; if I am despised by all the world I still give praise to God; and therefore I never met with a bad morning. You prayed that God would prosper me; to which I answered that I never was otherwise than prosperous; for, having learned to live with God, I know for certain that all he does must necessarily be for the best; and therefore whatever happens to me, by his will or his permission, whether it be pleasant or disagreeable, sweet or bitter, I always receive with joy as coming from his merciful hand, for the best, and therefore I never was otherwise than prosperous. You wished me also all happiness, and I, in like manner, replied that I had never been unhappy; for I have resolved to adhere to the divine will alone, and have so absolutely relinquished self-will as to will always whatever God wills, and therefore I was never unhappy; for I never desire to have any other will than his, and therefore I resign my will entirely to him.'

"Then said the Doctor, 'But what would you say if it should be the will of this Lord of majesty to cast you down into the bottomless pit? What would you do then?'

" 'How?' replied he hastily. 'Cast me down into the bottomless pit! His goodness holds him back therefrom. Yet if he should really do so I have two arms to embrace him withal. One arm is true humility, by which I am united to his most sacred humanity. The other is my right arm of love, by which I am united to his divinity. And with both I would embrace him so closely and hold him so firmly that he would be obliged to go down with me, and I would much rather choose to be in hell with God than in heaven without him.'

"Then understood the Doctor that true resignation to the divine will, accompanied with profound humility, is the shortest way to God. Having afterward asked the beggar whence he came, the latter replied, 'From God.'

" 'But where,' said the master, 'did you find God?'

" 'I found him,' said the other, 'where I forsook all creatures.'

" 'And where or with whom did you leave God?'

" 'I left him with the clean of heart, and amongst men of good will.'

" 'But I pray thee tell me who or what art thou!'

"And the beggar replied, 'I am a king. My kingdom is in my soul; for I can govern both my exterior and interior senses so absolutely that all the affections and forces of my soul are in perfect subjection to me; which kingdom is doubtless more excellent than all the kingdoms of this world.'

" 'What has brought you to this perfection?' inquired the Doctor.

"And the other answered, 'My silence, my heavenward thoughts, my union with God. For I could rest in nothing less than God. Now I have found my God I have everlasting peace and joy in him.' "*

"THEOLOGIA GERMANICA."

This is one of the few great devotional treatises of the world, setting forth, as its title-page says, "Many fair lineaments of divine truth, and very lofty and lovely things touching a perfect life." It was discovered by Luther and published by him, for the first time, in 1516. He says in his preface: "Next to the Bible and St. Augustine, no book hath ever come into my hands whence I have learned, or would wish to learn, more of what God and Christ and man and all things are." Luther esteemed Tauler to be its author. It is in his style and contains his sentiments, but it is now considered more probable that it originated at a little later date than Tauler's time, and was written by some other member of the class to which he belonged. It was the practice of these "Friends of God" to conceal their names as much as possible when they wrote, lest the desire for fame should mingle in their endeavors to be useful.

No fewer than seventeen editions appeared during

---

* We have followed mainly the version given in Francis of Sales's *Introduction to a Devout Life*, where it is taken from *The Works of J. Thaulerius, D.D.*, printed at Paris, 1623, and is called "A Conference on the Means of Attaining Christian Perfection." Whittier's poem "Tauler" is a description of the same incident.

the lifetime of Luther, and up to the present day it has continued to be the favorite handbook of devotion in Germany, as well as being widely circulated in other lands. Baron Bunsen says: "With Luther I rank this short treatise next to the Bible, but unlike him should place it before rather than after St. Augustine. This small but golden treatise has been now for almost forty years an unspeakable comfort to many of my Christian friends." Its main theme is self-renunciation, the laying aside of our own will in order to the accomplishment of the divine. It dwells upon the intimate union possible between God and man through love, enlightenment, the practice of virtue, and the cheerful endurance of trials. Charles Kingsley says: "To those who really hunger and thirst after righteousness, and therefore long to know what righteousness is that they may keep it; to those who long to be free from sin, and therefore wish to know what sin is that they may avoid it; to those who cannot help seeing that the doctrine of Christ in every man, as the indwelling Word of God, is a tenet which runs through the whole Bible, this noble little book will recommend itself."

The style of this treatise on German theology is quite mystical, and not many Americans of the present day, especially among the young, would be likely to read it through with much satisfaction. We append a few extracts, such as appear to us the most important from a practical point of view, and from these a fair idea of the character of the whole volume may be obtained:

"To learn an art which thou knowest not four things are needful. The first and most needful of all is a great desire and diligence, and constant endeavor, to learn the art. And where this is wanting the art will never be learned. The second is a copy or example by which thou mayest learn. The third is to give earnest heed to the master and watch how he worketh, and to be obedient to him in all things, and to trust him and follow him. The fourth is to put thine own hand to the work, and to practice it with all industry. But where one of these four is wanting the art will never be learned and mastered. So likewise is it with this preparation to be possessed with the Spirit of God."

"No one can be made perfect in a day. A man must begin by denying himself and willingly forsaking all things for God's sake, and must give up his own will, and all his natural inclinations, and separate and cleanse himself thoroughly from all sins and evil ways. After this let him humbly take up the cross and follow Christ."

"A true lover of God loveth him alike in having and in not having, in sweetness and bitterness, in good or evil report and the like, for he seeketh only the honor of God, and not his own, either in spiritual or natural things. Therefore he standeth alike unshaken in all things."

"All disobedience is contrary to God, and nothing else. In truth, no thing is contrary to God; no creature nor creature's work, nor anything that we can name or think of, is contrary to God, or displeasing to him, but only disobedience and the disobedient man. In short, all that is is well-pleasing and good in God's eyes, saving only the disobedient man."

"The man who is truly godlike complaineth of nothing, but of sin only. And sin is simply to desire or will anything otherwise than the one perfect good and the one eternal will, or to wish to have a will of one's own."

"Sin is to will, desire, or love otherwise than as God doth. Things do not thus will, desire, or love; therefore things are not evil, all things are good."

"Union with God is brought to pass in three ways, to wit, by pureness and singleness of heart, by love, and by the contemplation of God."

"Be assured he who helpeth a man to his own will helpeth him to the worst that he can."

"Time is a paradise and outer court of heaven, and therein there is only one tree forbidden, that is, self-will."

"There is nothing more precious to God or more profitable to man than humble obedience. In his eyes one good work wrought from true obedience is of more value than a hundred thousand wrought from self-will, contrary to obedience."

"He who is truly a virtuous man would not cease

to be so to gain the whole world; yea, he would rather die a miserable death. To him virtue is its own reward, and he is content therewith, and would take no treasure or riches in exchange for it."

"THE SPIRITUAL COMBAT."

The Spiritual Combat, which has for its motto the words of St. Paul, "A man is not crowned except he strive lawfully," was the production of Lorenzo Scupoli, an Italian monk of the order of the Theatines. He was born in the city of Otranto, about 1530, and died at Naples in 1610. After an active social life in populous cities he was driven into retirement by some shocking calumny, the exact nature of which is not known, and there, in quiet, patient meditation, this little book was born. It attained immediately an enduring popularity, and has been blessed to great multitudes of the choicest spirits of the earth. While the author yet lived it had been spread abroad in fifty editions and had been translated into many languages. In one hundred and ninety years there were two hundred and sixty editions, and all the tongues of Europe, as well as some in Asia, had received it. It was the favorite companion of Francis of Sales, of all human books his guide to holiness, doing more than anything else to mold and fashion that marvelous saint. He calls Scupoli "my master in all the exercises of the inward life." He carried the book in his pocket for eighteen years, reading daily some portion of it, and never re-reading it, he says, without profit. Its style is very simple and concise. It contains—including the supplements, wherein are "Maxims for the guidance of a soul that wishes to love Jesus Christ perfectly," and a treatise on "Inward Peace"—sixty-one short chapters. It is a capital manual for those who wish to make themselves masters in the art of godly living. The following selections are all we can make room for:

"I will give thee two rules, by observing which thou wilt live in this wicked world in as much quiet as possible. One is that thou strive with all diligence to close the door of thy heart more and more against desires. For desire is the upright beam of the cross, and of disquiet, which will be heavy in proportion to the greatness of the desire. And if the desires be many, many will be the beams prepared for many crosses. Then when difficulties and hindrances come, so that the desire is not fulfilled, behold the transverse beam, the cross of the cross, to which the desiring soul is nailed. Whoso, then, wishes not for the cross, let him give up the desire; for so soon as he gives it up he will have come down from the cross. There is no other remedy.

"The other rule is this: When thou art annoyed and offended by others, do not let thy mind dwell upon them, or on such thoughts as these: 'that they ought not so to have treated thee, who they are, or who they think themselves to be,' and the like. For all this is fuel, and a kindling of anger and wrath and hatred. But in such cases turn instantly to the strength and commands of God, that thou mayest know what thou oughtest to do, and that thy error be not greater than theirs."

"Everything which befalls us comes from God for our good, and we may profit by it. And though some of these (such as our own failings, or those of others) cannot be said to be of God, who willeth not sin, yet are they *from* him, in that he permits them, and though able to hinder them hinders them not."

"In all things make it a general rule to keep thy wishes so far removed from every other object that they may aim simply and solely at its true and only end, that is, the will of God. For in this way will they be well ordered and righteous; and thou, in any contrary event whatsoever, wilt be not only calm but contented; for, as nothing can happen without the Supreme Will, thou, by willing the same, wilt come at all times both to will and to have all that happens and all that thou desirest."

"As we should do our utmost to recover our peace of mind when we have lost it, so we must learn that there is nothing which ought to take it away or ever disturb it. Be assured that all disquiet is displeasing in his sight; for be it what it may it is never free from imperfection, and always springs from some evil root of self-love. For the disquiet thou feelest on account of thy sin comes not from having offended God, but from having injured thyself. If when thou fallest thou art so saddened and disquieted as to be tempted to despair of advancing and doing well, this is a sure sign that thou trustest in thyself and not in God. Consider that all these disquieting things and such like evils are not real evils, though outwardly they seem so, nor can they rob us of any real good, but are all ordered or permitted by God for righteous ends."

"Consider that not only do all the works which thou hast done fall short of the light which has been given thee to know them, and the grace to execute them, but also that they are very imperfect, and but too far removed from that pure intention, and due diligence and fervor, with which they should be done, and which should ever accompany them."

"The exercise of doing all things with the single aim of pleasing God alone seems hard at first, but will become plain and easy by practice, if with the warmest affections of the heart we desire God alone and long for him as our only and most perfect good."

"We are wont to pray most perfectly by placing ourselves silently in the presence of God, darting from time to time sighs unto him, turning our eyes to him with a heart longing to please him, and with a quick and burning desire that he would help us to love him purely, to honor and serve him."

"The aim of the whole life of the Christian who wills to become perfect must be a striving to form the habit of daily forgetting self more and more, and accustoming himself not to do his own will, that so he may do all things as moved thereto by the sole will of God, in order to please and honor him."

"Study to do some one act with as great fullness of will and purity of heart as if in it alone consisted all perfection, and the whole pleasure and honor of God."

"Happen what may, remain thou ever steadfast and joyful in humble submission to his divine providence."

"The key which unlocks the secrets of the spiritual treasury is the knowing how to deny thyself at all times and in all things."

"Purpose in all things to do what thou canst and oughtest to do; be indifferent and resigned to all that may follow out of thyself."

"Speak as little as may be of thy neighbor, or of anything that concerns him, unless an opportunity offers to say something good of him."

"Let everything be a means of leading thee to God, and let nothing hinder thee on the way."

"RELIGIO MEDICI."

Sir Thomas Browne, the author of *Religio Medici,* or *The Faith of a Physician,* was born in

London in 1605, and died at Norwich in 1682. He was knighted by King Charles II in 1671, on the occasion of his visit to Norwich. The little book which has chiefly perpetuated his name and fame, though he composed several others, was written about the year 1636, simply for his own satisfaction. The manuscript, however, was passed from hand to hand among his friends, and one of the many copies made was surreptitiously published in 1642. This compelled Dr. Browne to bring out an authorized and corrected edition in the following year, and a Latin version, issued in 1644, carried the name of the author throughout Europe with almost unparalleled rapidity, translations being at once made into French, German, Dutch, and Italian.

The book contains an account of the author's opinions and feelings on moral and religious subjects, and has been greatly admired and enjoyed by very many from that day to this. The style is strikingly original, and has a peculiar quaint eloquence which has commended it to multitudes. It breathes a noble charity and tender forbearance toward opponents, and can scarcely be read without profit, although it is rather speculative than spiritual, and is not very likely to be of much practical benefit to the ordinary mind. He was an earnest seeker for knowledge, with a vigorous, independent intellect, which caused him to be charged by some with skepticism. But these charges had small foundation.

He was a truly pious person and a sincere Christian, firmly attached to the Church of England. We append sufficient quotations to give the reader a little idea of the scope and quality of the book:

"I could never divide myself from any man upon the difference of an opinion, or be angry with his judgment for not agreeing with me in that from which within a few days I should dissent myself."

"At the sight of a cross or a crucifix I can dispense with my hat, but scarce with the thought or memory of my Saviour. I cannot laugh at, but rather pity, the fruitless journeys of pilgrims, or contemn the miserable condition of friars; for though misplaced in circumstances, there is something in it of devotion. I could never hear the Ave Maria bell without an elevation, or think it sufficient warrant, because they erred in one circumstance, for me to err in all, that is, in silence and dumb contempt; whilst therefore they direct their devotions to her, I offer mine to God, and rectify the errors of their prayers by rightly ordering mine own."

"When we desire to be informed, it is good to contest with men above ourselves; but to confirm and establish our opinions it is best to argue with judgments below our own, that the frequent spoils and victories over their reasons may settle in ourselves an esteem and confirmed opinion of our own."

"In expectation of a better I can with patience embrace this life, yet in my best meditations do often desire death. I honor any man that contemns it, nor can I highly love any that is afraid of it. For a pagan there may be some motives to be in love with life; but for a Christian to be amazed at death, I see not how he can escape this dilemma, that he is too sensible of this life or hopeless of the life to come."

"No man can justly censure or condemn another, because indeed no man truly knows another. . . . Further, no man can judge another, because no man knows himself; for we censure others but as they disagree from that humor which we fancy laudable in ourselves, and commend others but for that wherein they seem to quadrate and consent with us."

"It is a most unjust ambition to desire to engross the mercies of the Almighty, not to be content with the goods of mind without a possession of those of body or fortune; and it is an error worse than heresy to adore these complemental and circumstantial pieces of felicity, and undervalue those perfections and essential points of happiness wherein we resemble our Maker. To wiser desires it is satisfaction enough to deserve, though not to enjoy, the favors of fortune; let Providence provide for fools. It is not partiality but equity in God, who deals with us but as our natural parents: those that are able of body and mind he leaves to their deserts; to those of weaker merits he imparts a larger portion, and pieces out the defect of one by the access of the other."

"Let me not injure the felicity of others if I say I am as happy as any. *Ruat caelum, fiat voluntas tua* ('Though the heaven fall, let thy will be done'), salveth all; so that whatsoever happens it is but what our daily prayers desire. In brief, I am content, and what should Providence add more? Surely this is it we call happiness, and this do I enjoy."

"I can hold there is no such thing as injury; that, if there be, there is no such injury as revenge, and no such revenge as the contempt of an injury; that to hate another is to malign himself; that the truest way to love another is to despise ourselves."

"Bless me in this life with but peace of my conscience, command of my affections, the love of thyself and my dearest friends, and I shall be happy enough to pity Cæsar. These are, O Lord, the humble desires of my most reasonable ambition, and all I dare call happiness on earth; wherein I set no rule or limit to thy hand of providence; dispose of me according to the wisdom of thy pleasure. Thy will be done, though in my own undoing."

RUTHERFORD'S LETTERS.

The Letters of the Rev. Samuel Rutherford have long been a classic with the devout. Says Cecil: "Were truth the beam, I have no doubt that if

Homer and Virgil and Horace, and all that the world has agreed to idolize, were weighed against that book, they would be lighter than vanity." Rutherford was born in Roxburghshire, Scotland, about the year 1600. He took his degree of A.M. at Edinburgh in 1621, and for some years acted as professor of humanity there. In 1627 he was settled as pastor at Anworth in Kirkcudbrightshire. Here he labored faithfully for nine years, but saw very little result. In 1636 he published a theological work against the Arminians, which gained him great credit in some quarters; but it led to his being called before the High Commission Court, which deprived him of his ministerial office and banished him to Aberdeen. In this stronghold of episcopacy and Arminianism he stayed two years, and from this place two hundred and twenty of the three hundred and fifty-two letters which make up the unabridged collection were written. In 1638, the Covenant having again triumphed in the land, he hastened back to Anworth. But in the following year he was constrained by the opinion of his brethren to accept the chair of divinity in the University of St. Andrew's, where he spent the remainder of his life. He was one of the Scotch Commissioners to the Westminister Assembly, and had a leading hand in drawing up the Shorter Catechism. For a work in the defense of liberty, called *Lex Rex,* he was summoned in 1660 to answer before Parliament on the charge of high treason. But he was on his deathbed, beyond the reach of human oppression. His last words were: "Glory, glory dwelleth in Immanuel's land." He entered it March 20, 1661.

The letters, collected by one who went to the Assembly with him as his secretary, range in their dates from 1628 to 1661. They have been translated into several languages, and are greatly prized by those who seek to grow in holiness. Richard Baxter said of them: "Hold off the Bible, such a book the world never saw." Some of the expressions are very striking and live long in the reader's mind. But the book is very large (554 octavo pages), much of the matter is necessarily of only local interest or somewhat commonplace, and not many are likely to be attracted by it in these modern days. The selections we supply will give a fair idea of the fervent spirit of the writer and the peculiarities of his style:

"Welcome, welcome, sweet, sweet cross of Christ! I verily think that the chains of my Lord Jesus are all overlaid with pure gold, and that his cross is perfumed, and that it smelleth of Christ."

"I desire not to go on the lee side or sunny side of religion, to put truth betwixt me and a storm; my Saviour did not so for me, who in his suffering took the windy side of the hill."

"If ye were not strangers here, the dogs of the world would not bark at you."

"Verily I was a child before; all bygones are but bairns' play. I would I could begin to be a Christian in sad earnest."

"O to be dead to all things that are below Christ, were it even a created heaven and created grace! Holiness is not Christ, nor are the blossoms and flowers of the tree of life the tree itself."

"I never knew, by my nine years' preaching, so much of Christ's love as he has taught me in Aberdeen by six months' imprisonment. I charge you in Christ's name to help me to praise."

"Welcome, welcome, Jesus, what way soever thou comest, if we can get a sight of thee. And sure I am that it is better to be sick, providing Christ come to the bedside and draw by the curtains, and say: 'Courage! I am thy salvation!' than to enjoy health, being lusty and strong, and never to be visited of God."

"How sweet is the wind that bloweth out of the quarter where Christ is! Every day we may see some new thing in Christ; his love hath neither brim nor bottom. O that I had help to praise him! He knoweth that if my sufferings glorify his name, and encourage others to stand fast for the honor of our supreme lawgiver, Christ, my wages then are paid to the full."

"I have been much self-accused for not referring all to God as the last end; that I do not eat, drink, sleep, journey, think, and speak for God; that practice was so short and narrow, and light so long and broad."

"It is possible that the success answer not your desire in this worthy cause. What then? Duties are ours, but events are the Lord's."

"I have benefited by riding alone a long journey, in giving that time to prayer, by praying for others; for by making an errand to God for them I have gotten something for myself."

"I see that mortification, and to be crucified to the world, is not so highly accounted of by us as it should be. O, how heavenly a thing it is to be dead, and dumb, and deaf to this world's sweet music!"

"My faith hath no bed to sleep upon but Omnipotency."

"Let him make of me what he pleaseth; provided he make glory to himself out of me I care not. If my Lord would be pleased I should desire that some were dealt with for my return to Anworth; but if that never be I thank God. Anworth is not heaven, preaching is not Christ."

"O that the heaven, and the heaven of heavens, were paper, and the sea ink, and the multitude of mountains pens of brass, and I able to write that paper, within and without, full of the praises of my fairest, my dearest, my loveliest, my sweetest, my matchless, and my most peerless and marvelous well-beloved!"

"In your temptations run to the promises; they be

our Lord's branches hanging over the water, that our Lord's poor, half-drowned children may take a grip of them; if you let that grip go you will go to the bottom."

"Build your nest upon no tree here; for ye see God hath sold the forest to death; and every tree whereon we would rest is ready to be cut down, to the end that we might flee and mount up, and build upon the Rock, and dwell in the holes of the Rock."

"It is certain that this is not only good which the Lord hath done, but that it is best."

"I think that my love to Christ hath feet in abundance, and runneth swiftly to be at him, but it wanteth hands and fingers to apprehend him. I miss faith more than love or hunger."

"I am sure that the saints, at their best, are but strangers to the weight and worth of the incomparable sweetness of Christ. O, we love an unknown love when we love Christ. O black sun and moon, but O fair Lord Jesus. O black lilies and roses, but O fair, ever fair, Lord Jesus. O all fair things, black and deformed, without beauty, when ye are beside the fairest Lord Jesus."

"THE SAINT'S EVERLASTING REST."

This book is thought to have been read more widely, perhaps, than any other of the sort—except à Kempis and Bunyan. It has certainly done an immense amount of good, and of all the one hundred and sixty-eight different works credited to the author has most effectively perpetuated his fame. Richard Baxter (born at Rowton, 1615, dying at London, 1691) was one of the most celebrated non-conformist divines of England. His early ministry of sixteen years (1640-56) at Kidderminster accomplished great things for the renovation of the place. After leaving there he was in no one position for any great period, owing to the unsettled state of the country and the turbulence of the times. But he preached mostly in London, suffering a good deal of persecution on account of his political sentiments.

He is said to have written *The Saint's Everlasting Rest, or a Treatise of the Blessed State of the Saints in their Enjoyment of God in Heaven,* when far from home and without any book to consult but the Bible, and in such a low state of health as to be in constant expectation of death for many months. On the title-page of the original edition we find these words: "Written by the author for his own use in the time of his languishing, when God took him off from all public employment." At that time he is supposed to have been a little over thirty years of age. It was first published in 1650. Very many, some of them exceedingly distinguished and useful men, have ascribed their conversion to reading it. It must be said, however, that a considerable part of the treatise is not especially adapted to the furtherance of devotion, but is theological rather than practical. We make a few extracts from the fourth part, which is a work in itself, and the best portion. Baxter calls it "The Directory for the getting and keeping of the heart in heaven, by the diligent practice of that excellent, unknown duty of heavenly meditation; being the main thing intended by the author in writing this book, and to which all the rest is subservient."

"Let thy eternal rest be the subject of thy frequent serious discourse; especially with those that can speak from their hearts, and are seasoned themselves with a heavenly nature. It is great pity that Christians should ever meet together without some talk of their meeting in heaven, or of the way to it, before they part. It is pity so much time is spent in vain conversation and useless disputes, and not a serious word of heaven among them."

"Improve every object and every event to mind thy soul of its approaching rest. As all providences and creatures are means to our rest, so they point us to that as their end. O that Christians were skillful in this art! You can open your Bibles; learn to open the volumes of creation and providence to read there also of God and glory. Thus we might have a fuller taste of Christ and heaven in every common meal than most men have in a sacrament. If thou art weary with labor, let it make the thoughts of thy eternal rest more sweet. Is thy body refreshed with food or sleep? Remember the inconceivable refreshment with Christ. Thus every condition and creature affords us advantages for a heavenly life, if we have but hearts to improve them."

"A heavenly mind is the freest from sin, because it hath truer and livelier apprehensions of spiritual things. Is converse with wise and learned men the way to make one wise? Much more is converse with God. If travelers return home with wisdom and experience, how much more he that travels to heaven! If our bodies are suited to the air and climate we most live in, his understanding must be fuller of light who lives with the Father of light. A heavenly mind is also fortified against temptation, because the affections are thoroughly prepossessed with the high and holy delights of another world. He that loves most will most easily resist the motions of sin."

"The liveliest emblem of heaven that I know upon earth is when the people of God, in the deep sense of his excellency and bounty, from hearts abounding with love and joy, join together, both in heart and voice, in the cheerful and melodious singing of his praises."

"The things contained in heavenly rest are such as these: a ceasing from means of grace; a perfect freedom from all evils; the highest degree of the saint's personal perfection, both of body and soul; the nearest enjoyment of God the chief good; and a sweet and constant action of all the powers of body and soul in this enjoyment of God."

"The most difficult part of heavenly contemplation is to maintain a lively sense of heavenly things upon our hearts. It is easier merely to think of heaven a whole day than to be lively and affectionate in those thoughts a quarter of an hour."

"Hindrances to leading a heavenly life upon earth: living in any known sin; an earthly mind; the company of the ungodly; frequent disputes about lesser truths, and a religion that lies only in opinions; a proud and lofty spirit; a slothful spirit."

"THE NONSUCH PROFESSOR."

The full title of this remarkable book is, *The Nonsuch Professor in his Meridian Splendor; or, The Singular Actions of Sanctified Christians, laid open in Seven Sermons, at All-Hallow's Church, London Wall, by William Secker.* Of the author very little is known except that he was a dissenting minister of the seventeenth century who preached at Tewkesbury and at London. The book first appeared in 1660. It has been well styled "a breviary of religion," also "a beautiful little work, worth its weight in gold." It is marked by eminent spirituality and great concentration of thought. Some of the expressions are very quaint and pungent. The text of the entire discourse—for though called seven sermons it is printed as one—is, "What do ye more than others?" And its perusal is well calculated to stimulate Christian activity. We reproduce a few of the pearls found in this old casket, assuring the reader that there are plenty more there just as good:

"If the mercies of God be not loadstones to draw us to heaven they will be millstones to sink us to perdition."

"If the life of Christ be not your pattern the death of Christ will never be your pardon."

"Where self is the end of our actions Satan is the rewarder of them."

"As the shadow of the sun is largest when his beams are lowest, so we are always least when we make ourselves the greatest."

"How many professors are there who have light enough to know what should be done, but have not love enough to do what they know!"

"If the sun be eclipsed but one day it attracts more spectators than if it shone a whole year."

"The water *without* the ship may toss it, but it is the water *within* the ship which sinks it."

"To do evil for good is human corruption; to do good for good is civil retribution; but to do good for evil is Christian perfection."

"A covetous man is fretful because he has not so much as he desires; but a gracious man is thankful because he has more than he deserves."

"We do not sail to glory in the salt sea of our own tears, but in the red sea of a Redeemer's blood."

"We are so far from paying the utmost farthing that at the utmost we have not a farthing to pay."

"Our worldly comforts would be a sea to drown us if our crosses were not a plank to save us."

"If youth be sick of the *will-nots,* old age is in danger of dying of the *shall-nots.*"

"God hath a crown for the runner but a curse for the runaway."

"This is the day of God's long-suffering; but the judgment day will be the day of the sinner's long-suffering."

"All they who refuse and reject Christ as a refining fire must be obliged to meet and feel him as a consuming fire."

"If the night of death find thee graceless the day of judgment will find thee speechless."

"God has three sorts of servants in the world: some are slaves, and serve him from a principle of fear; others are hirelings, and serve him for the sake of wages; others are sons, and serve him under the influence of love."

"To bless God for mercies is the way to increase them. To bless God for miseries is the way to remove them."

"No Christian has so little from Christ but there is ground for praise; and no Christian has so much but he has need of prayer."

"By fasting the body learns to obey the soul; by praying the soul learns to command the body."

"Faith is the great receiver and love is the great disburser."

"The only way to keep our crowns on our heads is to cast them down at his feet."

"When once a man becomes a god to himself he then becomes a devil to others."

"It is better to lose the smiles of men than the souls of men."

"Reader, I would neither have you be idle in the means nor make an idol of the means."

"A man can never enjoy himself till he be brought to deny himself."

"The covenant of grace without us turns precepts into promises, but the spirit of grace within us turns promises into prayers."

"Good works may be our Jacob's staff to walk with on earth, yet they cannot be our Jacob's ladder to climb to heaven with. To lay the salve of our services upon the wound of our sins is as if a man who is stung by a wasp should wipe his face with a nettle, or as if a person should busy himself in supporting a tottering fabric with a burning firebrand. In proof of sanctification good works cannot be sufficiently magnified; but in point of justification good works cannot be sufficiently nullified. The lamp of duty can only shine clearly as it is trimmed with the oil of mercy."

"THE ART OF ALWAYS REJOICING."

Alphonsus de Sarasa, author of the book with the above title, was born of Spanish parentage in Flanders, 1618. He was a ripe scholar, a profound philosopher, and a great preacher. His many labors early consumed a feeble frame, so that he died at the age of forty-eight. But before departing he gave to the world, in 1664, a work which has well perpetuated his fame. Weigel, who translated it from the Latin into German, styles it "an incomparable and golden book." The distinguished Leibnitz gave it the highest praise. The large work, in fifteen treatises, is now very rare; but a compendium of it drawn up by the author himself, translated into English from the Italian version, was published in Boston twenty-five years ago. It is from this edition we make our extracts:

"As Epictetus well says, men are troubled not by things, but by the opinions they have about things. And the mischief of such ideas consists in this, that I wish to see everything done according to my fancy; and because this does not happen I am annoyed at everything. This is the one thing in the world which afflicts us, the sole wellspring of all our troubles, the opinion that nothing is done as it ought to be; by which we mean that nothing is done as we would have it. In order to reach peace we must correct this folly. What happens as we wish will make us most happy."

"The thought that God regulates all human events with infinite wisdom, that everything happens by the supreme design of God, is of the greatest use in composing the mind to peace. It is sufficient to know that all is governed by God, that 'nothing is done in the world of sense and sight,' as St. Augustine affirms, 'which is not by command or permission from that invisible court.' Nothing which takes place in nature happens by chance. And do not actions which proceed from the free will of man happen by God's disposing providence? We read in Scripture that, having often foreseen them, he described them to the prophets many years before, and they came to pass afterward exactly as he had predicted. How could he know of them so long before, and with such certainty, if chance and not his divine mind had directed them?"

"The providence of God, in order not to interfere in the least with man's free will, having foreseen in the immense volume of events, and well weighed, how each person would have acted under such or such circumstances, selected those circumstances and that position in which man could use his free choice in such a way that his free action should lead infallibly to that which God, in his wisdom, had foreordained. For if you look at the proximate cause leading to the result it may often appear to you to be chance; but if you wish to enter into the mind of God, who remotely disposes the said causes, you will understand clearly the deep counsel which produces that effect. I see it in the history of Joseph, as in that of many others."

"We ought to consider all well done which is done by God. Why should I feel disturbed about human events when I see infinite wisdom presiding over and ruling them? Am I so foolish as to believe that God does not know what is best to be done; or that, though knowing it, he does not wish to do it; or that, wishing it, he is not able to do it? Whatever may come I will certainly approve of it. Nor can I do better than spare myself the least doubt as to God's will being the best."

"God selects that state of life which is most suitable for each person. In no other state of life would my salvation be so secure, nor in any other state could I so well promote his glory. Whatever I am, I am from God; and only because I am from God I am what I am. And it is good for me to be thus; nor, if I could, would I wish to be otherwise than I am, for fear of opposing so much wisdom."

"He who is contented with his state of life ought also to be content with those things which led to it. Do not lose your peace if, after having made use of the means necessary for succeeding in your intention, it was not the will of God that you succeed; for, if he does not wish, though you were to move heaven and earth you could not even raise yourself a hand's breadth from your position. Nor, if you wish to live happily, must you compare the condition of others with your own. For if you wish to compare yourself with others you must weigh all the troubles of their state, nor wish to put them aside and only consider the happy side; then compare the blessings and the evils of your state with the blessings and the evils of those whom you envy, and you will see clearly that nothing is wanting to you, and that all has been dealt out to you justly."

"If we are pleased with our own state of life we ought to be pleased with everything belonging to it. I ought to be content with my poverty, and not wish to change it, because it is the state in which God wishes me to be; if I am satisfied with what I am, what can deprive me of peace?"

"We ought to be content with the annoyances incident to our state of life. Do you suppose that any state of life is without its peculiar trials and vexations? If so, you are mistaken. And it would be necessary for us not to be men, if we would not suffer calamity. He who wishes that the winds should not blow, or the waves be in motion, does not wish to sail, but to remain in the midst of the ocean without reaching the port. And what are evil tongues, evil-speaking, murmurings, calamities, and injurious words but winds which guide us to our desired haven? Evils would not disturb us were it not for the opinion that we have of evils, for we often think those things a hindrance which wonderfully assist us in our journey toward eternal happiness. How often by the very circumstances which I considered evils I have been led to that prosperity to which I should not have attained under more ordinary and peaceful circumstances!"

"Shall I wish to be otherwise than I am when by means of those very daily things which happen to me I am being conducted to eternal happiness? If you can say with the poet, 'Wish to be what you are, and wish nothing more than this,' you have found the short road to happiness, and also the only true joy of life. You can attain to this in any station, whatever it may be; and if you are content with your own because it pleases God to place you in it you are already happy."

"There is no other true happiness in the world except that of a soul content with its condition. This is the way to carry heaven about with you, and to be filled with the delights of paradise in this valley of tears. If you seek elsewhere for happiness you will seek in vain."

"Perfection consists in this, not only in bearing the changes of human fortune with patience, but in welcoming them and approving of them. This is true happiness, to wish things to be as they are, and not otherwise; this is the root of that grand 'Thy will be done,' by saying which we not only give God our will, but also our intellect."

"We must, in order to have always a right frame of mind, have a high conception of divine wisdom, for this is the foundation of all human tranquillity. Nor is it necessary for us to search into the reasons of everything in order to keep our mind calm and quiet; it is only necessary to believe firmly that nothing can take place in nature but what is ordered by the divine providence of God."

"Whatever happens to me, I will be on thy side, O my God, and will take thy part amongst men, and I will bravely affirm that all happens to me justly; for I shall ever be able to fight better when, lamenting my evil passions, I defend thy holy decrees."

"If, O reader, this divine sentiment is firmly rooted in your mind you are already happy and blessed; you rejoice in tribulation, because your faith sees clearly that those grievous things which you suffer are ordered by the wise providence of God, and you rejoice that they happen in order that God's divine will may be accomplished. This is the source of all joy. From this fountain springs that peace which overflows our heart and which keeps it at rest amidst the storms and turmoil of human events. He who attains to this breathes a pure air disquieted by no tempest; he has found the peace which the world cannot give, and which gives him happiness to the full."

"THE PRACTICE OF THE PRESENCE OF GOD."

The seventeenth century gave us, as we have already seen, Sir Thomas Browne, Rutherford, Baxter, Secker, and Sarasa. It gave us also a much less distinguished man than these, one who was altogether unlearned; who after having been a soldier and a footman was admitted as a lay brother among the barefooted Carmelites at Paris in 1666, and was afterward known by the appellation of Brother Lawrence, although Nicholas Herman was his original name. Converted at the age of eighteen, he walked before God on the earth until he was eighty, when he was received up. He only left behind him fifteen short letters, but their piety rescued them from oblivion; and prefixed to them are certain conversations with him written by one of his contemporaries and published at the instance of the Cardinal de Noailles. The substance of the ideas presented will be discovered in the following paragraphs:

"He told me that the foundation of the spiritual life in him had been a high notion and esteem of God in faith; which when he had once well conceived he had no other care at first but faithfully to reject every other thought, that he might perform all his actions for the love of God. That there needed neither art nor science for going to God, but only a heart resolutely determined to apply itself to nothing but him, or for his sake, and to love him only."

"He told me that all consists in one hearty renunciation of everything which we are sensible does not lead to God; that we might accustom ourselves to a continual conversation with him, with freedom and in simplicity. That we need only to recognize God intimately present with us, to address ourselves to him every moment. That the most excellent method he had found of going to God was that of doing our common business without any view of pleasing men, and (as far as we are capable) purely for the love of God. That his prayer was nothing else but a sense of the presence of God, his soul being at that time insensible to everything but divine love; he continued with God, praising and blessing him with all his might, so that he passed his life in continual joy. That we ought not to be weary of doing little things for the love of God, who regards not the greatness of the work, but the love with which it is performed."

"I have no will but that of God, which I endeavor to accomplish in all things, and to which I am so resigned that I would not take up a straw from the ground against his order, or from any other motive but purely that of love to him."

"I make it my business only to persevere in his holy presence, wherein I keep myself by a simple attention and a general fond regard for God, which I may call an actual presence of God; or, to speak better, an habitual, silent, and secret conversation of the soul with God. My continual care has been, for above forty years, to be always with God; and to do nothing, say nothing, and think nothing which may displease him; and this without any other view than purely for the love of him, and because he deserves infinitely more."

"Think of God the most you can. Let one accustom himself, by degrees, to this small but holy exercise; nobody perceives it, and nothing is easier than to repeat often in the day these little internal adorations. A little lifting up of the heart suffices; a little remembrance of God, one act of inward worship, are prayers which, however short, are very acceptable to God."

"There is not in all the world a kind of life more sweet and delightful than that of a continual conversation with God. For the right practice of it the heart must be empty of all other things. The presence of God is a subject which, in my opinion, contains the whole spiritual life, and whoever duly practices it will soon become spiritual."

"Let all our employment be to know God; the more one knows him the more one desires to know him. And as knowledge is commonly the measure of love, the deeper and more extensive our knowledge shall be, the greater will be our love; and if our love of God were great we should love him equally in pains and pleasures."

"SELF-RENUNCIATION."

The Abbé Guilloré, a contemporary of Fénelon and belonging to the same school of piety, lived just about two centuries ago. His monument is the treatise which he wrote on *Self-Renunciation; or, The Art of Dying to Self and Living for the Love of Jesus.* The book was composed, in French, in the form of "Conferences" addressed to a young friend under the author's instruction. Most of it is as well adapted to the Protestants of to-day as to the Roman Catholics of the past, for whom it was primarily prepared. Some of the topics taken up are: "Self-surrender the Only Path to Perfection;" "The Importance of Little Things;" "The Sensitiveness of the Holy Spirit;" "Half-hearted Service;" "The Interior Life of Jesus;" "Government of the Tongue;" "The Greatness of God's Mercy." The following extracts will give a taste of the quality of the work:

"God's glory and forgetfulness of self—such must be the aim of all true spiritual life. We offer up our life to God's glory when every action, however trifling, is performed for his sake. There is also a passive surrender to God, which lies chiefly in a loving acceptance of whatever he may lay upon us. He deigns to accept all, even our most trifling actions; nothing is too worthless to be offered to him, nothing is really unimportant, since we can serve him thereby. Be assured there is no happiness to be found on earth save in God, and in a complete loving surrender of self to him."

"There are three things which are the groundwork of all perfection, and which are attainable by all who will seek them steadfastly. These are, first, a calm exterior; second, a quiet heart; and, third, simplicity in our dealings with God. External composure is a great help to interior recollection. Of course it is true that a recollected mind tends to produce external tranquillity, but it is no less true that habitual external calmness and self-control do gradually promote interior recollection, and those who would lead a hidden life must cultivate a calm, unruffled demeanor in outward things. Watch the lives of those who are closely united to Jesus, and you will find that even externally they bear the signs of an indescribable calmness and peace which none else can know."

"An eager longing for success, or anxiety to prove our own wisdom and judgment, tends also to produce restlessness and perplexity of heart. Herein lies real peace of mind and true detachment. The soul that has learned to stay itself upon God does not care to risk the loss of such heavenly rest for the turmoil of this world's interests, and with the aid of his grace it fulfills all needful exterior avocations without being soiled or disturbed by their contact. Before you can acquire a thoroughly tranquil heart you must learn to care but little for the consequences, of what you do, leaving all such matters to God; laboring to the best of your ability for him, and being perfectly satisfied that he should grant success or failure as he sees fit."

"But one thing in this life is needful to you, that is, a heart stayed wholly on God."

"He never allows his creatures to exceed him in generosity. He appreciates your sacrifice, and will restore it fourfold, filling your soul with the gift of his own peace."

"It is a great mistake to fancy that attention to trifles in the spiritual life is unnecessary, or that God's glory is only promoted in great things; it is often harder to serve him well in seeming trifles than in those we call great. Real self-mortification is perpetual and knows no limit; its sincerity lies just in that very fact, and in the necessity for bringing every movement of the heart and of the body into captivity. If you would advance in true holiness you must aim steadily at perfection in little things, and beware of supposing that you seek God's glory in anything savoring of display and outward demonstration. Great works seen and known of men are too likely to carry the insidious poison of self-satisfaction in their rear, filling us with a pleasant impression of our own merits and importance as compared with others. But when a man is steadfast in conquering himself in little things, simply to please God, such a single aim, and the detachment which comes therewith, is a true offering to him, and surely promotes his glory."

"Depend upon it, a ready spirit of censoriousness is the rock on which many good men make shipwreck. Whenever it is possible, defend the absent, or, if that is impossible, turn the conversation."

"In the spiritual life one's sole aim should be to do all that depends upon ourselves, and then to bear patiently whatever depends upon God only. Those who have learned to wait patiently have made a vast stride in the spiritual life."

"It is a good rule in all we do to think less of the duty to be fulfilled than of how we may keep close to God while fulfilling it, so that our hearts may be more engrossed by him than our hands with work."

"Heedlessness and levity are flood gates through which spiritual blessings soon flow away, and the soul is left poor and barren."

"Habitual slackness is more destructive than casual acts of mortal sin; these last carry their own terror and warning, while the many trifling sins which accumulate where there is no effort to attain perfection do not startle the conscience, and often pass unnoticed."

"If our sufferings are caused by our fellow-men, how often we fail to look beyond the immediate cause to God, who is their real author, and in so doing turn such crosses to our own hurt, giving way to complaints, self-defense, or revenge, and calling our troubles hard and unjust."

"Suffering is inevitable; the question is, will you use it to your sanctification? It is a hard thing to suffer unprofitably when you have the power of turning all your crosses into blessings through that union with our dearest Lord which alone teaches us to lose ourselves in finding him. You cannot set aside the discipline; you may throw away all its healing grace."

"True obedience waits gently and without weariness, accepting what is in accordance with its own wishes, or the contrary, in the same trustful, patient spirit, having but the one aim—to please God. Lovingly accept whatever he may lay upon you."

"THE LOVE OF RELIGIOUS PERFECTION."

It was less than half a century ago, in 1851, that there appeared in Rome a treatise with the above title, which has passed through many editions in different places and has been translated into several languages. Some have compared it with *The Imitation of Christ* and *The Spiritual Combat,* to both of which it bears resemblance. The author was Joseph Bayma, of whom we know nothing except that, like Rodriguez, Guilloré, and Sarasa, he was a highly esteemed member of the Society of Jesus. He wrote the volume primarily for his own improvement, as an aid in carrying out the full idea of a religious life, dividing it into three books, which treat respectively of the motives, means, and exercise of virtue. Among many other excellent things he says:

"Whoever takes no care to advance has already begun to retreat, and become worse than he thinketh. If thou wilt preserve what thou hast, aim at what is more perfect."

"Let our study be to study what is more perfect. If we fail, let us be sorry for it; if we have an opportunity of practicing virtue, let us not pass it unheeded; let us take care to carry off each day some little victory over our vices."

"If thou be still solicitous about earthly goods, about the opinions of men, and worldly glory, behold thou hast not yet given thy whole heart to God, but kept it for thyself and the world."

"Meditation is the workshop of the spirit, the auxiliary of virtues, and the nursery of good works. It is the noblest exercise of self-denial, the torch of the mind, the life of the will, the bearer of divine grace, the anticipated likeness and imitation of the joys of heaven."

"Blessed is he that studies daily to know Christ more perfectly and advance in his love. The knowledge of Christ pours joy and sweetness into the soul, and renders the exercise of all virtues most easy."

"Thou shouldst care for nothing else in this world but to become daily more dear to Christ."

"He that knows but little cannot know how much remains for him to learn. But he that hath learned much knows so much the better how much remains yet to be learned by him. So they that are still full of passions and unmortified in their will often think that they have made sufficient progress; but holy and perfect men mourn, and think themselves very imperfect, for they see how much perfection they have still to acquire."

"Think not thyself holy, all at once, because thou dost foster holy desires; for it is one thing to desire and another thing to execute what is holy."

"If anything good befall thy brother, think it has fallen to thyself; be glad, and congratulate him from thy heart. If any evil, think it has happened to thyself; be sorry, and sympathize with him from thy very soul. If he seeks anything refuse him not; if anything annoys him, do it not; if he has formed a judgment or opinion about anything, resist it not. Be gentle, meek, polite, humble of heart; do not contend or murmur; ridicule not, satirize not, and, unless it be thy duty, reprehend not."

"Virtues are barely acquired after much labor, and are quickly lost by idleness."

"We know not whether God may not have decreed that on our progress should depend the salvation of many men, whose blood he will hereafter demand at our hands."

"O that thou wouldst frequently turn over in

mind the thought of a blessed eternity! Assuredly such a thought would excite thee to undergo labors, stimulate thee to abandon thine own ease, and urge thee to value nothing but virtue."

"Certainly pagans and infidels, and all that have no hope, may well be sad; but by what right is a servant of God overpowered with sadness in labors and crosses to which the kingdom of heaven is promised?"

"I call heaven and earth to witness that I had rather be a poor worm by the will of God than a seraph on high without it. I had rather, with the will of God, do nothing and be a martyr of idleness, than without it convert the whole world and be a martyr for the faith. I had rather, with the will of God, lie hidden in some wretched corner under a bushel than without it shine resplendent in the heavens. I had rather be a stock, with the will of God, than without it work miracles. Provided always I execute what is well pleasing in thy divine sight, wherever I am, whatever I do, I am quite great enough, quite rich enough, quite happy enough, quite wise enough."

A LIST OF TITLES.

FOR the convenience of the reader, and his assistance if disposed to procure for himself a set of these books that he may make his own selections, we append a list of titles, with publishers. Some of the volumes are no doubt out of print, and only to be picked up at secondhand stores. In the case of some, notably *The Imitation of Christ,* there is a vast variety of editions. No attempt has been made to catalogue these. It could not be done without an expenditure of time entirely out of proportion to any probable benefit that would be conferred. The authors are named here, as nearly as possible, in chronological order; only such authors and books being mentioned as are quoted from in the previous pages. The number of authors, it will be seen, is twenty, and the volumes about forty.

AUGUSTINE'S Confessions. James Parker & Co., Oxford and London, 1868. Pp. 248.
Rivington, London. 16mo.
Andover, 1860.

TAULER. Selections from the Life and Sermons of the Rev. Doctor John Tauler. Roberts Brothers, Boston, 1888. Pp. 155.

Theologia Germanica. Translated from the German by Susanna Winkworth. With a Preface by the Rev. Charles Kingsley, and an

Introduction by Prof. Calvin E. Stowe, D.D. W. F. Draper, Andover, Mass., and John P. Jewett, Boston, 1860. Pp. 275.

À KEMPIS. Imitation of Christ, by Thomas à Kempis. With an Introductory Essay by Thomas Chalmers, D.D., and a Life of the Author, by C. Ullmann, D.D. Gould & Lincoln, Boston, 1863. Pp. 283.

D. Lothrop & Co., Boston. Pp. 207.

—— An Extract of the Christian's Pattern; or, A Treatise on the Imitation of Christ, written in Latin by Thomas à Kempis. By Rev. John Wesley, A.M. Eaton & Mains, New York. 24 mo, pp. 196.

SCUPOLI. The Spiritual Combat. James Parker, Oxford and London. Pp. 242.

FRANCIS OF SALES. Introduction to a Devout Life. The Catholic Publication Society, New York, 1870. Pp. 396.

—— A Treatise on the Love of God. P. O'Shea, New York, 1868. Pp. 591.

—— Practical Piety. Webb & Levering, Louisville. Pp. 360.

RODRIGUEZ. Christian Perfection. Burns & Oates, London. 2 vols. Pp. 408, 373.

BROWNE. Religio Medici, by Sir Thomas Browne, M.D., with the Observations of Sir Kenelm Digby. Cassell & Co., New York. Pp. 192.

BAXTER. The Saint's Everlasting Rest; or, A

Treatise on the Blessed State of the Saints in Their Enjoyment of God in Heaven. By Richard Baxter. Abridged by Benjamin Fawcett. Worthington Company, New York, 1888. Pp. 297.

Taylor. Holy Living and Dying; with Prayers: containing the Complete Duty of a Christian. By the Rev. Jeremy Taylor, D.D. With a Memoir of the Author. D. Appleton & Co., New York, 1865. Pp. 535.

Rutherford. Letters of the Rev. Samuel Rutherford, with a Sketch of his Life, by the Rev. A. A. Bonar. Robert Carter and Brothers, New York, 1866. Pp. 554.

—— A Garden of Spices. Extracts from the Religious Letters of Rev. Samuel Rutherford. By Rev. L. R. Dunn. Eaton & Mains, New York.

Secker. The Nonsuch Professor in his Meridian Splendor; or, The Singular Actions of Sanctified Christians, laid open in Seven Sermons at All-Hallow's Church, London Wall. By William Secker. To which is added The Wedding Ring, a Sermon by the same author. With an introduction by C. P. Krauth, D.D. Sheldon & Co., New York, 1860. Pp. 320.

—— A String of Pearls from an Old Casket. P. E. Book Society, Philadelphia, 1860. Pp. 160.

SARASA. Compendium of the Art of Always Rejoicing. By F. Alphonsus de Sarasa. H. A. Young & Co., Boston, 1872. Pp. 140.

LAWRENCE. The Practice of the Presence of God the Best Rule of a Holy Life; being Conversations and Letters of Brother Lawrence. Willard Tract Repository, Boston. Pp. 67.

GUILLORÉ. Self-Renunciation. From the French of Guilloré. With an Introduction by the Rev. T. T. Carter. Rivingtons, London, Oxford, and Cambridge, 1871. Pp. 345.

FÉNELON. Christian Counsel on Divers Matters Pertaining to the Inner Life. G. W. McCalla, Philadelphia. Pp. 160.

—— Spiritual Letters. Same publisher. Pp. 56.

—— Selections from the Writings of Fénelon, with a Memoir of His Life. By Mrs. Follen. James Monroe & Co., Boston, 1858. Pp. 374.

BAYMA. The Love of Religious Perfection; or, How to Awaken, Increase, and Preserve It in the Religious Soul. By Father Joseph Bayma. John Murphy & Co., Baltimore, 1865. Pp. 254.

UPHAM. Principles of the Interior or Hidden Life, designed particularly for the consideration of those who are seeking Assurance of Faith and Perfect Love. By Thomas C. Upham. D. S. King, Boston, 1843. Pp. 462.

UPHAM. A Treatise on Divine Union; designed to point out some of the Intimate Relations between God and Man in the higher forms of Religious Experience. H. V. Degen, Boston, 1851. Pp. 435.

—— Life of Faith. Harper & Brothers, New York, 1864. Pp. 480.

—— Life of Madame Catharine Adorna; including some leading facts and traits in her religious experience, together with explanations and remarks tending to illustrate the doctrine of Holiness. Harper & Brothers, 1864. Pp. 249.

—— Life and Religious Opinions and Experience of Madame de la Mothe Guyon; together with some account of the personal history and religious opinions of Fénelon, Archbishop of Cambray. Harper & Brothers, New York, 1874. Two vols., pp. 431, 377.

FABER. A Sketch of his Life, together with Selections from his Devotional Works in Poetry and Prose, by Rev. James Mudge. Christian Witness Company, Boston, 1885. Pp. 264.

—— Spiritual Conferences. By F. W. Faber, D.D. John Murphy & Co., Baltimore, 1867. Pp. 472.

—— Growth in Holiness. By F. W. Faber, D.D. Murphy & Co., Baltimore, 1866. Pp. 494. (There are also six other prose works of Faber's, published by Murphy, whose titles are given on a previous page. And there are many editions of or selections from his poems. There is an unabridged edition of the Hymns, pp. 427, published by H. H. Richardson & Co., New York, and Thomas Richardson & Son, London.)

GOULBURN. Thoughts on Personal Religion; being a Treatise on the Christian Life in its two chief elements, Devotion and Practice. D. Appleton & Co., New York, 1866. Pp. 428.

—— Pursuit of Holiness; a sequel to Thoughts on Personal Religion. Intended to carry the reader somewhat further onward in the Spiritual Life. D. Appleton & Co., New York, 1870. Pp. 261.

—— The Idle Word; Short Religious Essays upon the Gift of Speech, and Its Employment in Conversation. D. Appleton & Co., New York, 1866. Pp. 208.

—— An Introduction to the Devotional Study of the Holy Scriptures. D. Appleton & Co., New York, 1866. Pp. 193.

INDEX

I

J

K

L

M

O

P

Q

R

S

T

www.ingramcontent.com/pod-product-compliance
Lightning Source LLC
Chambersburg PA
CBHW021501210326
41599CB00012B/1082

* 9 7 8 3 7 4 3 3 2 4 5 3 4 *